光明城
LUMINOCITY

看见我们的未来

Conference Proceedings of the International Committee of Architectural Critics (CICA)

国际建筑评论家委员会研讨会论文集

Local Architecture Strategies in the Context of Globalization

Edited by
Li Xiangning
Manuel Cuadra

李翔宁
〔秘〕曼努埃尔·夸德拉
主编

全球化进程中的地方性建筑策略

本书受国家自然科学基金面上项目"基于关联域批评话语分析的当代中国建筑国际评价认知模式与传播机制研究"（51878451）资助

上海·同济大学出版社　TONGJI UNIVERSITY PRESS

目 录
CONTENTS

08 序　写给国际建筑评论家委员会研讨会 | 王澍
09 Preface　A Message to the International Committee of
　 Architectural Critics Conference | Wang Shu

10 前言　超越东南西北：当代建筑中的普遍性和特殊性 | [秘] 曼努埃尔·夸德拉
11 Preface　Beyond Orient and Occident, North and South: The Universal and the
　 Particular in Contemporary Architecture | Manuel Cuadra

18 梁思成与丹下健三 | [日] 矶崎新
19 Liang Sicheng and Kenzo Tange | Arata Isozaki

32 建筑评论的向度 | 李翔宁，邓圆也
33 Dimensions of Architectural Criticism | Li Xiangning, Deng Yuanye

48 认同——人之境况的建筑：由五个篇章和一个尾声构成的思考 | [秘] 曼努埃尔·夸德拉
49 Identity — The Architecture of the Human Condition: A Reflection in Five Movements
　 and One Epilogue | Manuel Cuadra

72 建筑及其调解的力量 | [德] 阿克塞尔·索瓦
73 Architecture and the Power of Mediation | Axel Sowa

92 自然和文化的共生：土耳其博德鲁姆德米尔别墅群 | [土耳其] 森古·奥依门·古尔
93 Nature and Culture Symbiosis: Demir Houses, Bodrum, Turkey | Sengül Öymen Gür

112 约翰内斯堡的普遍性与特殊性：迈向再生性发展 | [南非] 凯伦·艾克尔
113 The Universal and the Particular in Johannesburg: Moving towards
　　Regenerative Development | Karen Eicker

132 当代拉丁美洲建筑：常识与理想主义 | [巴西] 露斯·沃德·采恩
133 Contemporary Architecture in Latin America:
　　Common Sense and Idealism | Ruth Verde Zein

148 主编简介
149 Brief Introduction of the Editors

序

写给国际建筑评论家委员会研讨会

王澍

因为这次会议，来自世界各大洲的一群建筑评论家和建筑师汇聚在同济大学和中国美术学院，讨论的焦点是全球化进程与地方性建筑策略的关系，这个话题可能是现在全球建筑界讨论的话题中非常让人期待的话题，也是真正的全球性话题。在我看来，当下建筑界的热点话题，比如流行的数码设计、智能生态建筑等等，更多是全球市场的烧钱玩具，是非常西方中心化的话题，并不是真正的全球话题。而讨论全球化进程中的地方性建筑策略，它触及的核心问题其实是我们如何面对"现代性"，这是全球话题，但它却是单向性的，因为它潜含的意思是：世界建筑的主流话语要以西方为参照中心。涉及地方性工作的建筑师不管借用多少地方元素，总是强调自己的工作是现代的，否则就好像没有归属或脱离了时代，这似乎是让建筑师非常恐慌的。但是，在真正的实践中，现代性从来不是一个抽象话题。例如，传统中国建筑主要结构是木结构，这其中至少一半会使用到夯土墙体，但目前国家颁布的技术规范完全没有关于这类木结构和夯土墙体的规范，这就使得一个伟大的建筑传统不可避免走向灭绝。问题是，这个传统不是考古学意义上的传统，它实际上还在中国的许多地方被广泛使用着。所以，是否现代只是一个表面问题，真正的问题是专业化与非专业化的建筑活动的冲突，专业化试图消灭的不是非专业化，它消灭的是真正的地方性和多样性。以我的多年实践体会，真正的地方性建筑策略，恐怕更多地发生在业余建筑工匠的手中。

Preface

A Message to the International Committee of Architectural Critics Conference

Wang Shu
Translation: Yang Liu (from Chinese to English)
Proofreading: Deng Yuanye

A group of architecture critics and architects from around the world have gathered here at Tongji University and the China Academy of Art to take part in this conference. Our discussions will focus on the relationship between globalization and vernacular architecture strategies – perhaps the most anticipated topic discussed in architecture circles worldwide, and undoubtedly a global theme. To me, the current hot topics in architecture, such as digital design, eco-smart buildings and other fashionable trends, are simply toys for the global market to throw money into; these are extremely Western-centric topics, not truly global ones. But the discussion of vernacular architecture strategies in the context of globalization touches on a core issue: how we should face "modernity". This is a global topic, but it is a one-sided topic as well, because its underlying meaning is that world architecture's mainstream discourse must look to the West. Architects involved in local projects will always emphasize the fact that their work is "modern", regardless of how many elements they may have borrowed from elsewhere, almost as if without modernity they can't be a part of something or will fall behind the times. This seems to have driven architects to panic. However, in its actual implementation, modernity has never been an abstract concept. For example, traditional Chinese architecture tends primarily to use wood structures, and no less than half use rammed-earth walls, but the technical specifications the government has released to date do not include any specifications related to wood structures or rammed-earth walls. As a result, the extinction of a great architectural tradition has practically been guaranteed. The problem is that these are not traditions in the archaeological sense, because in reality they are still seen in widespread use in many regions throughout China. Thus, modernity or lack thereof is a superficial issue. The core issue is the conflict between specialized and non-specialized architectural activities – specialization seeks to eliminate non-specialization, and in doing so eliminates true localness and diversity. Based on my many years of practical experience, I'm afraid that real vernacular architecture strategies are more commonly found in the hands of amateur architectural artisans.

前言

超越东南西北：
当代建筑中的普遍性和特殊性

[秘]曼努埃尔·夸德拉

译者：邓圆也（英译中）

20世纪是非常重大的复兴时期。这一变化的基础是由工业化决定的。工业化始于18世纪的英国和法国，这两个国家通过它们的殖民地，获取并拥有它们所必需的资本、原材料和巨大的市场。19世纪下半叶，工业化的影响扩展到了其他国家。尤其在德国，工业化进程迅速推进，使它成为英国和法国的竞争对手。然而，君主制的德国在政治和社会方面仍然落后。

第一次世界大战是一个转折点：出现了历史上第一个在德国领土上建立的共和国，以及一个民主系统和一个社会福利系统。在建筑学领域，这一时期兴起的"现代运动"将自己理解为这一转变的一部分，即工业化产生的新的条件下的自然结果，以及一个正在有意地创造自己的存在的工业社会。这个时代的建筑师非常有意识地让建筑从过去的遗迹中解放出来，赋予它新的根基——更贴近新时期经济、社会、文化和艺术特色的根基。

所谓"现代运动"实际上是一系列实验性的平行变化。功能主义关注的是工人、雇员等新用户，新的用途以及其对建筑物空间、形态组织和设计的影响。这就相当于建构主义如何把兴趣点放在探索新建筑系统和结构的建筑维度，或者理性主义与空间和物体的智识发展的形式，有机主义与物体本身想拥有的完形，表现主义与材料和结构、颜色和纹理、形式和空间的情感和感官维度。

在20世纪20年代，世界用赞赏的眼光观看这一巨大的创造力爆发，但对其社会和政治背景知之甚少。"国际风格"一词，源于美国，却影响到了全世界，它总结了这些不同趋势和特点，并认为一种"超级文化"或者"超国建筑"正在萌芽。同时，人们用"国际风格"谈论风格的结果是将发展的地位降低到一个形式的问题，让其变成不过是种种历史对话中的另一个声音，这在19世纪以及20世纪上半叶保守圈子里是很常见的。

Preface

Beyond Orient and Occident, North and South: The Universal and the Particular in Contemporary Architecture

Manuel Cuadra

The 20th century was a time of a very fundamental renewal. The basis for these changes had been set by the process of industrialization. Industrialization had its beginnings in 18th century England and France, two countries whose colonies afforded them the necessary capital, raw materials, and huge markets. In the second half of the 19th century, industrialization began to spread to other countries. Germany, in particular, underwent a very rapid transformation which made the country competitive with England and France. However, Germany was still ruled by a monarchy, and remained politically and socially backward.

The First World War was a turning point for Germany: a republic was established for the first time on German soil, and both a democratic system and a welfare state emerged. In architecture, the Modern Movement during this period understood itself to be a part of these transformations, that is, a consequence of the new conditions generated by industrialization, and an expression of an industrial society that was deliberately creating its own existence. The architects of this time very consciously pursued the liberation of architecture from the relics of the past in order to give it a new foundation – one in accordance with the economic, social, cultural, and artistic characteristics of the new age.

The Modern Movement was, in fact, a series of parallel developments that were basically experimental in nature. The functionalists were concerned with new users – workers, employees – and new uses, as well as their implications for the spatial and formal organization and design of buildings. This was similar to how constructivists were concerned with exploring the architectural dimension of new building systems and structures, rationalists with the intellectual development of the forms given to spaces and objects, organicists with determin-

第二次世界大战后在世界范围内传播的，更多的是一种偏离了欧洲风格本质的现代主义风格，与这一时期激进的政治、社会和文化动荡毫无关系。20世纪50年代的现代主义以一种美式的、关于现代性作为美国人生活方式的一部分的理解为特征，其基础是美国看似非政治性的自由市场经济，并与这个时期的大众社会和它的消费主义有关，即以大规模生产、标准化以及统一性为基础。

上述风格的不同趋势出现在了一些欧洲国家，例如北欧国家，以及日本和美国本身。然而，它们的存在主要是地方性的，并且，许多情况下，是个别卓越者的功劳。从全球来看，建筑师建造的建筑质量在操作层面上，大多是一个市场问题，包括房地产市场、金融市场、建造市场以及建筑市场。事后看来，可以称之为一种"邪恶联盟"，即新鲜的、社会和政治方面均激进的现代性连同其在资本主义制度下产生的成果的商业开发。

从历史上看，目标在于重定建筑方向的后现代运动的起源之地正是现代运动诞生之地，但不同的是，它并不抵制19世纪关于城市以及城市建筑的理念，而是从中学习。批判性地域主义者为了克服后现代运动典型的历史决定论，强调了地方，即物理维度中的地方、文化维度中的地方以及地方作为在历史上业已发展的、应当在现代背景中重新诠释的地域性建筑形式的基础。由于自由市场经济和以消费主义为基础的美国生活方式继续存在，导致该后现代趋势产生的成果停留在形式的、肤浅的表面。

自20世纪90年代以来，美国这种生活方式在全球迅速扩展，特别是在中国、巴西、俄罗斯、印度、南非等新兴经济体，而这些都是人口庞大的国家和市场。同时，美国生活方式和涡轮资本主义被忽略的一面，包括其对环境、社会以及全球共存的影响，只是慢慢地才被人们所意识到。

20世纪70年代罗马俱乐部第一次指出了这些危险，其范围从极大的能源消耗、对化石燃料的依赖以及伴生的大气变暖、不受控制的气候变化，到社会逐渐分化为一小群得益于这一发展的精英、一个艰难追赶日益增长的消费社会生活标准的中产阶级阶层，以及远远未能过上有尊严生活的广泛的社会阶层。

向中国、印度等国家的发展扩散正在深刻地改变这些国家自身以及整个世界。争夺资源和市场的竞争日益加剧，自然资源的稀缺性锐化了冲突。我们正在接近罗马俱乐部宣称的转折点，在此转折点之前，科学和技术的进步能够弥补按照美国的生活方式发展所带来的不利之处，而在此之后不再可能。因此争夺资源的斗争将变得更加尖锐——也就是说将发生国家之间的公开冲突和与自然的斗争。

在这种背景下，国际建筑评论家委员会（CICA）的成员在上海聚集，我们都深刻意识到了目前建筑领域的特殊情况以及可能产生的影响。我们以各大洲的发展作为深思的出发

ing the gestalt that things themselves desired, and expressionists with the emotional and sensual dimensions of materials and structures, colors and textures, and forms and spaces.

During the 1920s, the world observed this overwhelming explosion of creativity with admiration and, at the same time, with little understanding of its social and political background. The term "International Style", which was coined in the USA but had worldwide repercussions, summarized these different tendencies and characteristics, and assumed the emergence of a "super culture" or "super national architecture". At the same time, when used in discussions of style, the term "International Style" reduced development to a formal question and made it just another conversation among many historical dialogues, as was typical in the 19th century and in conservative circles in the first half of the 20th century.

What spread around the world after the Second World War was more of a modernistic style, one that had moved away from its European roots and had little to do with this period of radical political, social, and cultural upheaval. The Modernism of the 1950s was characterized by the US-American understanding of modernity as part of an American way of life practiced in the US. It was based on the country's apparently apolitical free market economy, and was related to the mass society of this period and its consumerism, that is, it was founded on mass production, standardization, uniformity.

Divergent tendencies appeared in several European countries, such as the Nordic states, in Japan, and in the US itself. However, such tendencies were primarily local and in many cases the achievements of outstanding individuals. Considered globally, the bulk of architecture made by architects on a practical level was a matter of the markets – the real estate market, the financial market, the building market, and the architecture market. In hindsight, one might say that there was an unholy alliance between an original, socially and politically progressive modernity and the commercial exploitation of the achievements of this modernity within a capitalist system.

Historically, the reorientation of architecture sought by the Postmodern Movement was to begin from the same starting point as the Modern Movement, but, rather than fighting against 19th century ideas of the city and city architecture, it hoped to learn from them. Critical Regionalism attempted to overcome the historicism typical of postmodern approaches by addressing the place – the place in its physical dimension, the place in its cultural dimension, and the place as the foundation of historically-developed regional architectural forms that required reinterpretation in a modern context. The continued existence of the free market economy and the consumerism-based American way of life meant that the consequences of

点。来自非洲、亚洲、大洋洲、欧洲、北美洲和南美洲的同事报告了一些典型的、他们认为是理解本国建筑必须知道的情况。结果是一个万花筒，迥异的场景并列，乍一看几乎无法比较，因为它们涉及非常不同的时刻。同时，所有这些场景都是同一个不可分割的现实的切面。

如今，我们非常清楚地认识到，构成这个世界的不同国家、民族和文化的发展是不能分开的。我们也知道，每一个不同的现实本身对于我们所生活的世界是必不可少的。我们今天的世界不可能如一些人在20世纪初所想象的那样，直接简化成某一个共同点。我们涉足的是影响着我们所有人的全球结构。与此同时，我们正在面对的社会、政治和文化多样性与这个世界的地理、气候和自然多样性一样丰富。

通过对具体情况的思考，我们的交流有助于使多样性有形化，而详细研究这些具体情况将澄清我们的世界多样性的基础究竟是什么，以及我们能够在每一个地方启发多少文化、艺术和建筑。此种启发来自人以及地方本身的特点：地方的地理学和气候学特点，如其植物和动物群落，这是最广泛意义上的物质维度；人的文化和社会特点，即组成社会的不同群体，他们有自己的历史，有特殊的传统，关于当下、现代性、工业化、消费主义，他们也有自己的观点，简而言之，对于构成人类生活的一切和那些不能简单概括的东西，他们都有自己的看法。特别与特殊无处不在，而特殊性——我们最不能因为强加全球统一的生活方式而使它变得枯燥无趣。

同时，普遍性也无处不在。我们很快意识到，凡是建筑师们工作的地方，无论是实践上活跃的建筑师，还是理论上积极的建筑师，我们所接触到的专家们在解决现实问题时都有着非常相似的途径和方法，并以非常相似的形式开发自己的项目——如建筑项目、城市项目、景观项目等——或者开发自己的理论和思考。因此，我们认识到，这个世界的建筑师、建筑史学家、建筑理论家，以及建筑评论家都能够畅通无阻地沟通和理解对方，无论他们来自哪里以及他们搜寻的方向和他们想实现的目标是什么。

因此，对我们来说，互相尊重也不是难事。我们认识到，表面上可能会把我们分开的东西，也是连接我们彼此的东西；我们知道这些不同只不过是人的不同侧面，这些侧面我们作为个体可能无法从自己的生活中接触到，而知道他人是谁、他人正在我们的世界里做什么能让我们更了解这些侧面。因此，我们认识到差异性不仅意味着利益冲突，同时也意味着相互充实以及更多。差异性，即个性，正如群体性、社会性一样，属于人类常态。

由此来看，寻求对话成为最佳和唯一途径，能协调利益并确保每个人最终能从共存中受益。没有必要共享同样的一种"国际风格"或"全球风格"。我们在今天的全球化时代所需要的亲密是人类之间的亲近，这一人类的亲近来自我们所共享的人类常态，而且必然包括个性，也因此包括特殊性。这二者必须如普遍性那样被很好地培养，因为它们都是卓有成效的共存

these postmodern tendencies remained purely formal, superficial, and cosmetic.

Since the last decade of the 20th century, the world has experienced a rapid expansion of this American way of life, especially in emerging countries such as China, Brazil, Russia, India, and South Africa – all countries (and markets) with very large populations. At the same time, consciousness of overlooked aspects of the American way of life and turbo capitalism, including its consequences for the environment, for society, and for global coexistence, has grown quite slowly.

The Club of Rome first pointed out these dangers in the 1970s, which ranged from enormous energy consumption, dependence on fossil fuels, and an associated warming of the atmosphere and uncontrolled climate change, to the polarization of society into a small group of elites benefiting from development, a middle class struggling to keep up with the rising living standards of consumer society, and a broad social strata far removed from a life of dignity.

The spread of these developments to China, India, and other countries is changing, in a profound way, both the countries themselves and the rest of the world. Competition for resources and markets is increasing, and scarcity of vital natural resources fuels conflicts. We are nearing the turning point announced by the Club of Rome, which marks the transition from the stage where scientific and technological progress can compensate for the disadvantages of development following the pattern of the American way of life, to a stage where this is no longer possible, and where, therefore, the struggle for resources will intensify – meaning open conflicts between countries and war against nature.

It was under these circumstances that the members of the International Committee of Architectural Critics (CICA) met in Shanghai, all of us very aware of the particular conditions and implications of architectural production today. We began by considering the developments on all continents. Our colleagues from Africa, Asia, Australia, Europe, Anglo-America, and Latin America reported on characteristic situations that they felt were essential to understanding the architecture of their respective countries. The result was a kaleidoscope, a juxtaposition of very different scenes, which at first glance can hardly be compared because they relate to very different moments. At the same time, all these scenes are simply different facets of a single, indivisible reality.

We are very much aware today that the developments in all the different countries and nations and cultures that make up this world cannot be separated. We also know that each of these different realities is, in and of itself, essential to the world in which we live. Our world

的基础，不仅是我们各社区卓有成效的共存基础，也是构成我们世界的各国和文化的卓有成效的巨大共存的基础——这共存超越了历史上划分的东西方世界，也超越了20世纪以来划分的南北世界。

today cannot simply be reduced to a common denominator, as imagined by some at the beginning of the 20th century. We are dealing with global structures that affect us all. At the same time, we are dealing with a social, political, and cultural diversity that is as rich as the geographic, climatic, and natural diversity of our world.

Our exchange served to make diversity tangible by contemplating concrete situations that, when considered in detail, make clear what exactly the diversity of our world is based on and how much cultural, artistic, and architectural production can be inspired in every place. This inspiration comes from the people and the characteristics of the place itself: geographic and climatic aspects of the place, such as its flora and fauna, which is its material dimension in the widest sense; cultural and social aspects of the people, comprising the different groups that make up society, groups who have their own history and particular traditions, their own view of the present, of modernity, industrialization, and consumerism, in short, their own view on all that constitutes human life and what cannot be simply generalized. The special and the particular can be found everywhere, and above all the particular is the one thing we should not make flat by imposing a globally uniform way of life.

At the same time, the universal is everywhere as well. We soon realized, that, wherever architects are working – practically active architects, theoretically active architects – we are dealing with experts who employ very similar approaches and methods in managing their realities, and who choose very similar forms when developing their projects – architectural projects, urban projects, landscape projects – as well as their theories and reflections. Thus we recognize that the architects of this world, as well as the architectural historians, architectural theorists, and architectural critics, are able to communicate and understand each other very well, no matter where they come from or in which direction their searches and their realizations are aimed.

For this reason, it is easy for us to respect each other as well. We recognize that what might superficially separate us is also what connects us to one another; we recognize these differences as no more than different facets of being human, facets to which we as individuals may have no access in our own lives, facets that we understand all the better because of what others are and what others are doing in our world. We thus recognize that differentness can also mean conflicting interests, but at the same time means mutual enrichment and much more. Being different, that is, individuality, is no less a part of the human condition than being in the community, being social.

It follows then that seeking dialogue is the best and only way to coordinate interests in order to allow everyone to ultimately benefit from our coexistence. There is no need to share a common "international style" or "global style". The closeness we need in today's globalized age is a closeness of human beings, a human closeness that arises from the common human condition we all share, and which necessarily includes individuality and thus the particular. Both of these must be cultivated with the same intensity as the universal, for they are the basis of a fruitful coexistence between our communities and a great coexistence between the nations and cultures that make up our world – a coexistence that goes beyond Occident and Orient, the historical subdivision of our world, and beyond North and South, the 20th century subdivision of our world.

梁思成与丹下健三

[日]矶崎新
译者：辛梦瑶，王西（日译中）

1930年代—1940年代

对日本来说，1945年，第二次世界大战终结，社会制度和生活意识发生重大变革，被认为是建筑史和文化史上的一道分水岭。然而，从所谓"近代建筑"的框架来看，显然1935年是更加根本的转换点。这是由于此时国际主义发生了全面的倒退。无论是有着包豪斯的德国，还是有着先锋的构成主义的俄国，都开始了统合社会主义和民族主义的讨论。然而，成功将这两者充满魅力地统合起来的只有墨西哥的现代主义者。而在日本，我认为这种统合大概是发生在1955年左右吧。依存于技术因而具有全球普遍性的近代建筑，不得不将曾经被排除在外的民族、传统、风土、遗产等作为设计的主题重新吸收进来。

1935年，日本的丹下健三是东京帝国大学工学部建筑学科的学生，此时的他仍然是一个对勒·柯布西耶充满崇敬的纯正的现代主义者。然而，一到了1940年代，他就成为一个打着"日本主义者"旗号的反动的激进分子。他"叛逃"了。

而在同时期的中国，在国民党的"围剿"之下，共产党的中央红军离开了井冈山，开始了"长征"，并且在遵义会议上，毛泽东掌握了实权。在此时期，从宾夕法尼亚大学毕业回国的梁思成先是与林徽因一道开设了事务所，而后成为了国民党政权设立的国家文物研究所的职员，开始了对古建筑的广泛调查。由于日军的侵略，国民党政府由南京迁往重庆之时，该研究所也迁往四川。而这些工作成果被整理成为《中国建筑史》。梁思成在美国接受的是被称为"布扎体系"的现代主义教育。这在他回国之后立刻着手设计的北京大学女子宿舍（1930年）中一目了然。古建筑的测绘成果是在1944年左右整理出来的，尽管在战乱之中并没有发表的机会，但在参加联合国大厦的设计讨论的前后，他在美国的大学所进行的讲座英文讲义则被作为 A Pictorial History of Chinese Architecture（《图像中国建筑史》）整理出来，并在

Liang Sicheng and Kenzo Tange

Arata Isozaki
Translation: Yang Liu, Deng Yuanye (from Chinese to English)

1930s–1940s

In 1945, the end of World War II brought remarkable changing of social constitution and life ideology, which marked the boundary between architectural history and cultural history to Japanese society. However, in the view of the framework of "contemporary architecture", the year of 1935 is a more essential turning point, for internationalism had an overall recession. All nations, not only Germany who has Bauhaus but also Russia who has avant-gardism, started a discussion on uniting socialism and nationalism. Unfortunately, only the modernism from Mexico united both of them. In Japan, I think the uniting happened around 1955. Due to dependency on techniques, modern architecture represented their global universality, but now we have to resume the design themes that were once excluded again, like nationality, tradition, customs and heritage.

In 1935, Kenzo Tange was still a student in Architecture Department of Engineering Faculty of Imperial University of Tokyo. At that time he was a pure modernist who showed great respect to Le Corbusier. However, once it reached 1940, he became a reactive radical who declared he was "Japanist". He "betrayed".

During the same period in China, under the "encirclement and suppress" of the Kuomintang, the central communist red army left Jinggangshan and started the Long March. At the Zunyi meeting, Mao Zedong took over the ruling power. Meanwhile, Liang Sicheng, who had graduated from University of Pennsylvania and returned to China, firstly opened an office with Lin Huiyin; later, he became a researcher of State Antique Research Center which was set

1980年代出版。

梁思成在1949年中华人民共和国成立后，进行了包括国徽、人民英雄纪念碑等数个国家象征物的创作。所以，他的《中国建筑史》很重要。而丹下健三在谈论广岛和平公园的纪念碑时，作为纯粹现代主义者盛赞勒·柯布西耶的言论，与作为日本主义者在1940年前后所提出的"日本国民建筑样式"的言论显然是一种参照。与此相似，梁思成的《中国建筑史》也成了最重要的参照物。

我并没有资格去评判梁思成关于中国建筑史的技术内容的正确与否，只是尽力地读了英文版罢了。其中所刊登的图片，都仅是图面化的梁思成本人的调查成果，以及在现场支着三脚架所拍摄的照片。不过，从梁思成这些直接面对遗迹和建筑物进行田野调查记录之后所描绘出的建筑图面中，我们就能感受到他所接受过的美国布扎风的图面表现的教育。从清末开始到中华民国初期，中国的文物研究一直被称为是日本研究者的专属舞台，日本人在中国各地进行了实测调查，也完成了诸多的研究报告。不过，看来似乎梁思成并没有参照那些报告。当时大概也得不到日语的报告吧。梁思成所绘山西省大同善化寺大雄宝殿和佛宫寺释迦塔的图面，可谓是建筑图面的杰作。在日本也有学习学院派风格图面的建筑师的作品，但那些不能与梁思成的作品相提并论。

可以说，梁思成以学院派现代主义建筑师的身份，对中国的历史建筑物进行了观察记录。正如上文所述，我认为在1940年左右，丹下健三由向往国际建筑的纯正的现代主义者转变为了"日本主义者"，但由于没有物证，也只是推测罢了。关于梁思成，虽然同样也只是推测，但他从美国刚刚回国之时所进行的学院派的现代主义的建筑设计，于1930年代中期以后则转向中国各地的历史建造物的调查，再之后，便是将根据地移向四川之后开始起草《中国建筑史》，这与丹下的转变是相似的。当梁思成在对古建筑的实物进行调查时，丹下正在将中国的建筑史作为"中国性的事物"（「中国的なもの」）的历史进行书写。

梁思成的《中国建筑史》的第一章题为"中国建筑之特征"。我认为，在日本，堀口舍己所提出的"日本性的事物"（日本的なもの）是同样的问题架构。堀口舍己一边批判着也被学院派称作"殖民地"式的"帝冠样式"，一边思考着"日本性的事物"。另一方面，在中国南京中山陵那样的"帝冠样式"基本上就是"国家样式"。在这种情况下，梁思成是从对实物的实地调查出发，忘却现代主义，以《营造法式》为范本，来完成那些记录。当然，在那个时候，比对维特鲁威以来的西欧建筑的基本形式，他将中国建筑用斗拱/柱/柱础这样的order来理解。在介绍《营造法式》的时候，他也提到源自希腊的柱式的收分做法，在这本宋代的著作中也有记载。

我所关注的是梁思成在书籍最开始的地方，便做出了"历用构架制之结构原则：既以木材为主，此结构原则乃为'梁柱式建筑'之'构架制'"的论断。从这一论断中，我们可以理解到

up by Kuomintang regime and started extensive investigation on ancient architecture. Because of Japanese invasion, the research center moved to Sichuan when the Kuomintang government moved its capital to Chongqing from Nanking. His investigation accomplishments were organized and came up to *The Architecture History of China*. In America, Liang Sicheng was educated under a modernism system which is considered as "The Beaux-Arts Tradition". And it was reflected in his work, Girl's Dormitory in Peiking University (1930), shortly after he returned to China. The fruitful mapping work was organized around 1944, despite of lacking opportunities to publish during war period, sometime before or after participating the design discussion on United Nation Building, his English presentation prepared for the lecture in American Universities was organized, i.e. *A Pictorial History of Chinese Architecture* and published in 1980.

After the foundation of People's Republic of China in 1949, Liang Sicheng undertook several "national symbols" designs such as national emblem, the monument and so on. Therefore, his work *The Architecture History of China* is very important. When Kenzo Tange was talking about the monument in Peace Park in Hiroshima, his reverent appraise on Le Corbusier as a "simply modernist" then and his remark on "Japanese Country Architecture Design" given around 1940 were apparently a comparison. Similarly, *The Architecture History of China* by Liang Sicheng also became the most critical reference.

Technically, I am not qualified to judge the correctness in "The Architecture History of China" by Liang Sicheng, but I try my best to read the English version. All published architecture pictures are the investigation drawings conducted by Liang Sicheng and photos taken on site with a tripod. Nevertheless, we can percept the Beaux-Arts Tradition education he received in America from all these architectural pictures he recorded while facing heritage and architecture and doing field investigation. From the end of Qing Dynasty to the beginning of Republic of China (1912-1949), the study on Chinese antique was dominated by Japanese researchers. Japanese conducted field surveying and mapping in different regions of China, and accomplished many research reports. However, it seems that Liang Sicheng did not refer to those reports. It was probable that he could not obtain Japanese reports at that time. Liang's drawings of the Great Shrine Hall of Shanhua Temple in Datong, Shanxi and Sakya Pagoda of Fogong Temple are considered masterpieces of architectural drawings. There are some architects who learn academic style drawing, however, their works are by no means at the same level as those of Liang Sicheng's.

It is fair to say that Liang Sicheng observed and recorded historic architecture of China as an academic modernist and architect. As stated above, I think Kenzo Tange turned into a

梁思成是将中国建筑看作"构筑的"。建筑评论家浜口隆一在《国民建筑样式论》中，将西欧与日本建筑特性的区别整理为：西洋＝"物体的·构筑的"/日本＝"空间的·行为的"。如果按照这样的区分标准来看，梁思成所指出的"中国性的特征"则可以说是西欧的特点。此外，书开头处所刊登的表现了从柱础到斗拱的中国的柱式图也十分引人注目。梁思成写到，与社会等级相对应，柱式的用法也是有限定的。这种视角应该是参考了清代的《工程作法则例》而并不是宋代的《营造法式》，不过由于梁思成所讨论的建筑是"官式"（包含佛寺）的建筑类型，所以也是理所当然的了。建筑是从属于社会制度的。

浜口隆一分别将"物体的·构筑的"、"空间的·行为的"总结为西欧的建筑和日本的建筑的特征。基于阿尔伯蒂以来的西洋建筑论的系谱，梁思成将架构性置于"中国性的"核心位置，是可以理解的。与此对应，作为"日本性的"日本建筑的特征是以怎样的根据得出的呢？作为先行研究，我想举出建筑史家足立康于1933年所进行的所谓"间面记法"的日本建筑的记述法的解读，以及堀口舍己于1932年写成的《茶室的思想背景及其构成》。这两位与梁思成在年纪上也几乎相差无几。

日本的建筑物从古便有所谓"三面四间"的说法，但到了明治时期，"间"、"面"分别指的是什么变得不明起来，而古文献的阅读上也经常出现误解。不过，足立康按照近代之后引入日本的西洋建筑史观，证明了所谓"间"是指正面的柱间的数目，而"面"则指的是建筑附加的"庇"在侧面的数目。在维特鲁威的《建筑十书》中，柱子的直径是按照一定的模数，与"柱间"形成一定的比例（symmetria），所以"间"作为一种比例，表达了被称为"开间"的正面的柱跨数。然而，进深方向由于大多总是两跨，所以被省略了，变为了用"面"去衡量母屋上到底附加了怎样尺度的"庇（ひさし）"。在中国，一般并不存在"庇"，可以说所谓"庇"的存在并不是正统的，而是十分"和样化"的用法。所谓"间面记法"的表述法，是用来描述包含"柱间"和"庇"的建筑所拥有的作为"空洞"的空间的。而浜口隆一所指出的传统的日本建筑的空间特性在于"空间的"理由也在于此。

在建筑史方法上给予足立康的"间面记法"以革命性之评价的，是东京大学的建筑史学家太田博太郎。另外，太田也评价道，在堀口舍己的著作《草庭》的解说中，尽管茶室已经消失了，但堀口舍己一边从追溯建筑史中的五官茶会记的细节开始，将茶人们的行为加入到想象性复原的思考方式中，通过审视在那个场所中设置的道具的配置，以及茶人活动的姿态，来实现空间的再现。这大概是将建筑空间作为"行为"来进行认知的日本性的建筑特性的含义吧。

浜口隆一将日本传统建筑定性为"空间的·行为的"建筑的论文，是在1944年发表的。这正值梁思成的《中国建筑史》的草稿完成之年。而浜口隆一的该理论，也让丹下健三参照伊势神宫和京都御所设计并得到好评的"大东亚建筑纪念营造计划"以及"盘谷日本文化会馆计划"在得到"日本国民建筑样式"的认定上变得具有正当性。

"Japanist" from a pure modernist who had a yearning for international architecture. But this is just a guess since there is no material evidence available, the same as that of Liang Sicheng. Nevertheless, Liang created the architecture design of academic modernism shortly after he had come back from the U.S.A, and then switched to different regions of China to investigate historic architecture after the middle period of 1930s, and started to draft *The Architecture History of China* after transferring base area to Sichuan. Kenzo Tange experienced akin changes as Liang Sicheng. When Liang Sicheng was investigating "real ancient architecture", he was writing the architecture history of China at the same time, taking it as the history of "the phenomena and things of China" (「中国的なもの」).

The first chapter of *The Architecture History of China* by Liang Sicheng is "the characteristics of architecture in China". In my opinion, "the phenomena and things in Japan" (日本的なもの) proposed by Horiguchi Sutemi has the same question framework as that of Liang Sicheng's. Horiguchi Sutemi criticized the "imperial crown design" i.e. "colony" style by the academics, and meanwhile, he reflected on "the phenomena and things in Japan". On the other hand, "the imperial crown design" likes Zhongshan Mausoleum in Nanking is basically the "state style" in China. For this reason, Liang Sicheng started investigation from "real stuff", abandoned modernism, and took *Architecture Rules* as a sample to complete those records. Of course, comparing to the basic forms of western architecture since M.Vitruvii Pollionis at that time, he understood architecture of China in such "order" as Dougong, Pillar, and Pillar Base. The proportioned columns' Greek Order origination recorded in the work of Song Dynasty was also mentioned when Liang Sicheng was introducing *Architecture Rules*.

At the very beginning of *The Architecture History of China* by Liang Sicheng where I paid the most attention to, he concluded that the structure principles of all framing systems in history, mainly using wood, are "pillar order architecture". From this conclusion we understand that Liang Sicheng deemed architecture of China as "structural". Architecture commentator Ryuichi Hamaguchi distinguished the characteristics between architecture in Western Europe and that of Japan in his work i.e. *The Discussion of Country Architecture Design* (「国民建築様式の諸問題」,1944), in which he says, Western architecture = "material-oriented and structural" while Japanese architecture= "spatial and behavior-oriented". According to this classification standard, "the characteristics of Chinese phenomena and things" indicated by Liang Sicheng can be categorized as western features. Moreover, the Chinese order pictures of pillar base to Dougong on the first several pages attract great attention. In the book, it says the application of order must follow rules which are corresponding to social class. This prospective should be inferred from *Architecture Building Standard and Sample* of Qing Period instead of *Architecture Rules* of Song Period. Nevertheless, the architecture types Liang Sicheng

1950 年代

梁思成与丹下健三都是在学习了西欧的现代主义建筑后又经历了战争，各自领悟出"中国式"和"日本式"的手法并加以运用，分别在1950年的中国和战后日本设计了国家的象征。人民英雄纪念碑是模仿石碑的塔状建筑，而广岛慰灵碑是模仿家型埴轮的包覆状建筑。前者既有方尖碑风格的轮廓，又保留了源于佛塔的中国式细部，同时毛泽东手书的宋体书法也浮于其上。后者强调了马鞍般的曲线，但刻于慰灵碑上的文字却拙劣得令人无法直视。两者虽有纪念碑与慰灵碑的区别，但同样作为祭奠逝者的设施，在经历全球形势不断变化的半个多世纪后，现在仍然作为国家的象征存在着。

两人都既是建筑师又是城市规划师，对接手过大型项目的他们来说这可以称得上是小项目。尽管在设计过程中都遭到了政治的蹂躏，但建成的作品可以称得上是失败的，只占他们作品集中的极小一部分。话虽如此，这些却是整个20世纪关于"中国式"与"日本式"的问题架构中不可或缺的案例。

中华人民共和国宣言在北京天安门城楼上被公布后，北京作为"首都"的基本重建方案开始制定，当时首先是由来自苏联的专家带领制定的。方案很显然参照了1930年代斯大林用七个大型尖塔构成的首都莫斯科，以及在中心高耸着列宁像的苏维埃宫。站在北京天安门城楼上，毛泽东说道："从这里看过去尽是城墙，我更希望看到冒黑烟的林立烟囱……"后来由周恩来向建筑师和城市规划师下达了这个指示。毁掉城墙再推进工业化的方案与苏联专家的近代化方针也吻合。另一方面在广岛，核武器破坏力量被重新解释为"对核的和平利用"，以此为名义的《和平宣言》诞生，这使得广岛成为用和平重新解读战争的模范城市。在毁灭于原子弹爆炸的广岛，能证明这座"城市"的慰灵碑是一个国家象征的就只有这份《和平宣言》。

在如火如荼的国共内战之后，人民英雄纪念碑受到已成功进行核开发的苏维埃联邦的支持而诞生，而毁灭于原子弹爆炸的城市在遭到攻占和统治之后被重新定义为"和平城市广岛"，这促成了凭吊原子弹爆炸受害者的慰灵碑。这些事件被各个建筑的主持建筑师们反复置于设计中。假设这些建筑对建筑师来说是失败的作品，他们将其以某个形象建成，我想是因为建筑师已经非常熟悉在模仿公认的民族文化遗产的基础上进行近代化设计的手法。

梁思成在1953年10月发表了《建筑艺术中社会主义现实主义和民族遗产的学习与运用的问题》。他正是在那年夏天刚刚访问完苏联回国。尽管斯大林在那年春天逝世，但梁讨论的是诞生于莫斯科建设过程中的"社会主义现实主义和民族遗产"的辩证统一。

在日本，同样的主题被称作"传统与创造"。丹下健三在这个时期设计了一系列广岛和平公园相关设施。爆炸十周年时设施一建成，他就在第二年发表了论文《现代建筑的创造与日本建筑的传统》（「現代建築の創造と日本建築の伝統」，1956年6月）。这些立刻遭到了建筑评

discussed are "official style" (including Buddhist temples). Therefore, his conclusion is merely natural—architecture is subordinated to social constitution.

Ryuichi Hamaguchi concluded the characteristics of architecture in Western Europe and Japan as "material-oriented and structural" and "spatial and behavior-oriented" respectively. Based on classification and categorization of western architecture theory since Alberti period, Liang Sicheng placed "framework" at the center of "Chinese feature", which is reasonable. Comparably, how did we come up to the conclusion of "Japanese feature" of Japanese architecture? As the background study, I would like to explain the so-called "Jian Mian Descriptive Method" which was a Japanese architecture descriptive method conducted by architectural historian Adachi Yasushi in 1933 and "The Cultural Background and Structure of Tea House" by Horiguchi Sutemi in 1932. Both of them are of similar age as Liang Sicheng.

Since ancient times, the characteristic of Japanese architecture had been described as so-called "Three Mian Four Jian". However, the definition of "Jian" and "Mian" became vague when it came to Meiji period, and mistakes appeared frequently in ancient books and documents. Despite that, according to the western architecture history theory introduced into Japan after modern times, Adachi Yasushi proved that "Jian" refers to the number of pillar intervals on the front; while "Mian" refers to the number of shelters affiliated to the architecture on the side. In *Ten Books on Architecture* by M.Vitruvii Pollionis, the diameter of pillar, according to certain module, forms a fixed symmetria with "pillar intervals". Therefore, "Jian" as a symmetria, shows "the number of pillar intervals" on the front which is called "length". However, the width is often omitted for two pillars are commonly seen on both sides, which is replaced with "Mian" to measure the size of "shelter （ひさし）" that affiliates to master room at the center. In China, there is no "shelter", in other words, it is not formal expression but a very Japanese style expression. The so-called "Jian Mian Descriptive Method" is used to describe architecture that have "pillars" and "shelters" as "hollow" space. That is why the space characteristics of traditional Japanese architecture are pointed out to be "spatial" by Ryuichi Hamaguchi.

The scholar who commented "Jian Mian Descriptive Method" by Adachi Yasushi as revolutionary one on architecture history was architecture historian Oota Hirotarou. In addition, Oota Hirotarou also remarked that, in *Kusaniwa* by Horiguchi Sutemi, despite that tea house has disappeared, Horiguchi Sutemi started from the detail of tea meeting recorded in five officers of architecture history, immersed the behaviors of tea-drinkers into imaginary complex thinking. Through reviewing the facilities on that occasion and the activities of the tea drinkers, he managed to reproduce the space. This is probably the meaning of perceiving

论家的批判，丹下也用《民众与建筑》《日本建筑师》(「民衆と建築」「日本の建築家」)两篇论文来回击。之后这段论战被整理为"传统论""民众论""职能论"，这个构成与毛泽东的著作《矛盾论》《实践论》《抗日游击战争的战略问题》相似。毛泽东将马克思置于中国革命战略的核心，与此相同，丹下健三将受到维也纳美术史学派影响的日本与斯大林时代的社会主义现实主义进行对照，制定了日本近代建筑发展方针。然而梁思成在那之后遭到了政治批判，与斯大林遭受赫鲁晓夫批判的逻辑一致。这是因为民族主义性设计与"大屋顶派"被视作同一事物。

1959年中国建成了"北京十大建筑"，在日本丹下健三设计建造了代代木体育场（1964年）。1950年时作为国家象征受到瞩目的纪念碑设计，十年后在两个国家都以大型城市建筑物的姿态示人。

第二次世界大战结束前，现代主义建筑向全球扩散，与各个地方被认为具有民族性与传统性的本土建筑相混杂，这个过程中争议频出。尽管每个地方都试图追溯古典式起源，但近代化仍在进行。因此虽然说是向古典回归，但一边运用以技术为基础的现代主义、一边以"中国式"和"日本式"为问题架构的建筑手法也逐渐形成。梁思成学习宋代建筑工程领域的专著《营造法式》，关注建筑的形式性与结构性，而丹下健三关注贯穿建筑内部（比例）与外部（城市）的空间性。

自近代国家建立以来，"（艺术）＝建筑＝城市＝国家＝（民众）"这个图式成为"理所当然的道理"，因此梁思成和丹下健三作为技术官僚建筑师参与制定国家性建筑的设计与制度，在1950年代被卷入了政治论战。两人都在无法逃避的国际政治形势中成为关注焦点。中国共产党与赫鲁晓夫保持对立路线，在政治与文化上采取了与苏联绝缘的独立道路，随后走向了"大跃进"和"文化大革命"的"失去的十年"。日本虽然通过《旧金山对日和平条约》（San Francisco Peace Treaty, 1954）重新获得独立统治权，但秘密签订与美军结盟的安全保障条约引发了日本国内阻止其签订的抗议运动，这个请愿示威被镇压后日本接受了美国的"核之伞"保护，随后开始了被称为"奇迹般的"1960年代经济高速增长。日本和中国与冷战时期的两极化大国关系发生变化，为后来的梁思成和丹下健三带去影响。梁思成经历了批判、反批判和自我批判后，在1972年文化大革命尚未结束时就去世了。丹下健三在东京奥运会和大阪世博会等国家重要事项上被委以重任，作为技术官僚建筑师完成任务。

梁思成和丹下健三都学习过国际建筑样式，也各自学习和阐释了自己国家的传统遗产。两人正因为建造了可成为国家象征的建筑，政治意识形态就成为不可避免的问题。然而人民英雄纪念碑与广岛慰灵碑各自通过模仿石碑和埴轮，在错综复杂的巨大压力下得以保持一种形态。在我看来，这是因为他们两人都对"（艺术）＝建筑＝城市＝国家＝（民众）"这个理念的技术主义建筑怀有信念。在无论从职业角度还是思考形式上都与这个等式完全脱离的今天来看，只有与新中国革命或原子弹爆炸导致的城市毁灭这样的历史性"切断"时刻碰撞后

Japanese style architecture characteristics as "behavior-oriented".

The architecture essay by Ryuichi Hamaguchi which defined Japanese traditional architecture as "spatial and behavior-oriented" was published in 1944 when Liang Sicheng completed the draft of *The Architecture History of China*. Ryuichi Hamaguchi's theory facilitated the reasonability of Kenzo Tange's highly appraised "Japanese Country Architecture Design" and "Japanese Cultural Center Plan in Bangkok" in recognition, which were designed for "Grand East Asia Architecture Memorial Center Construction Plan" with reference to Naign Shrine and Kyoto Imperial Palace Park.

1950s

Liang Sicheng and Kenzo Tange, both of them had studied modernist architecture of Western Europe and experienced war, had worked out "Chinese Style" and "Japanese Style" designing method independently, and furthermore, applied them when they were designing national emblem for new China in 1950 and for Japan after war respectively. The People's Heroes Monument is a tower-shape imitation of stone tablet while Hiroshima Peace Memorial Monument is an imitation of Haniwa shelter-like architecture. The former has preserved both the shape of obelisk style and delicacy work of Buda tower of Chinese style. What's more, the handwriting calligraphy in Song typeface of Mao Zedong is relieved on it. While the latter emphasizes saddle-like curve, but, unfortunately the writings carved on it is extremely artless. Though they are monuments for different purpose – one for heroes' memorial and the other for resting victims in peace, both of them as facilities to pray for the dead remain the state symbol until present after experiencing continuous global changing for half a century.

As both architects and urban planners, designing the monuments was just a small project to them since they had undertaken big projects before. Though they had experienced political tormenting during designing process, their works which were considered failure only take up a very small proportion. Despite that, these are also indispensable samples in the question framework about "Chinese style" and "Japanese Style" in the whole 20th century.

After the announcement of new China declaration was published on the Tian'anmen Rostrum, the reconstruction plan "capital" of Beijing was initiated. At that time, the first group to make the plan were led by experts from the Soviet Union. The plan was apparently referred to the Palace of Soviets whose center was set up with high-rise Lenin sculpture and

的"匹夫之勇"才能支撑那个信念。之后梁思成在文革的批判中去世，丹下健三离开日本成为中东国家的皇室建筑师，社会也变得不再需要曾经的官僚主义建筑师了。

capital Moscow which was built with seven huge spires by Stalin in 1930. Standing on the Tian'anmen Rostrum in Beijing, Mao Zedong said, "I see nothing other than rampart from here. I prefer to see numerous chimneys emitting black smoke." This direction was given to architects and urban planners by Enlai Zhou. The plan which suggested to destroy rampart and progress industrialization also matched the principal of modernization of experts from Soviets. On the other hand, in Hiroshima, nuclear destructive power was reinterpreted as "use for peace", and "Peace Declaration" was generated in this name, which made Hiroshima a model city that reinterprets war with peace. In Hiroshima, the city that was destroyed in the atomic bombing, "Peace Declaration" is the only evidence to prove that this "City" monument is a national symbol.

Among the fierce civil war between the Kuomintang and the Communist Party in Republic of China, People's Heroes Monument was built with the support of Soviet Federal who had successfully developed nuclear power; while the city that was destroyed by atomic bomb was reinterpreted as "peace city Hiroshima". Where it was actually being conquered and ruled, the monument facilitated for victims died in atomic bomb. These events have been used repeatedly in designing various architecture by architects in chief. Assuming that these architecture are failure works of architects and they built them like some certain images,, I think it is because these architects are very familiar with the method of modernization design by imitating recognized national cultural heritage.

"The Questions of Learning and Applying Socialism Realism and National Heritage in Architecture Art" was published in October, 1953 by Liang Sicheng. Back then he had just finished visit to Soviets and returned to China in summer. Despite that Stalin passed away in spring of the same year, but Liang was proposing a discussion on the dialectical unity of "socialism realism and national heritage" which was generated in the process of Moscow construction.

The same topic was discussed in Japan with a different name "the tradition and creation". In this period, Kenzo Tange designed a series of facilities affiliated to Hiroshima Peace Park. One year after decennial architecture of the A-bomb was finished, he published an essay i.e. "The Creation of Modern Architecture and Architecture Tradition in Japan"（「現代建築の創造と日本建築の伝統」, June, 1956）. These were criticized by architecture critics immediately and Kenzo Tange debated them with two essays i.e. "The Folk and Architecture" and "Japanese Architects"（「民衆と建築」「日本の建築家」）. Afterwards, these debates were organized as "Tradition Theory", "Folk Theory", "Function Theory", whose structures were similar to the great works of Mao Zedong, i.e. "Contradiction Theory", "Theory of Practice" and "Discussion on Guerrilla Warfare". Mao Zedong placed Marx at the center of the revolutionary strategy of China. Similarly, Kenzo Tange compared The Vienna School of Art History which influenced "Japanism" with socialism realism in the time of Stalin and established modern architecture development policy of Japan. Unfortunately, Liang Sicheng encountered political criticism, like Stalin was repudiated by Khrushchev, because nationalist design was considered to be the same thing as "Big Roof Supporters" (i.e. formism).

In 1959, "Ten Great Buildings in Beijing" were built in China; while Yoyogi Stadium (1964) was designed and built by Kenzo Tange in Japan. Both of the monument designs, which received great attention in 1950 as national symbols, were posed as mega city architecture after ten years in their own countries respectively.

Before the end of World War II, modernist architecture spread the whole world and mixed with local architecture which were considered as national or traditional by the natives. Contradictions arouse during this process. Despite that all regions were trying to track the classic and traditional originality, modernization kept ongoing. Therefore, modernism depending on techniques and architecture building methods based on the framework of "Chinese style" and "Japanese Style" were gradually forming despite that people tended to return to classic and traditional. When Liang Sicheng studied *Architecture Rules* of Song Dynasty, he paid attention to the format and structure of architecture. While Kenzo Tange focused on the spatiality which integrated interior (symmetria) and exterior (urban) spaces.

Ever since the foundation of modern country, the equation of "art=architecture=city=state=people" had become "more than logical", therefore, Sichewng Liang and Kenzo Tange participated in designing state architecture design and making constitution as technique bureaucrat architects. And this was why they were involved in the political controversy. Both of them stood in spotlight because of the inescapable international political situation. The Communist Party of China contradicted with Khrushchev and chose to be independent on politics and culture, and afterwards experienced "the big leap" and "the lost 10 years" of culture revolution. Though Japan had regained its independent authority through "San Francisco Peace Treaty, 1954", its secret tie with American army i.e. the security assurance treaty arose protestors event in Japan. After being suppressed by the government, Japan had accepted the protection of "Nuclear Umbrella" offered by the U.S.A. Since then, Japan experienced the "incredible" economy growth in 1960s. Both Japan and China had dramatic changes in the relationship with the two polarized big countries during the Cold War, which brought impact to Liang Sicheng and Kenzo Tange afterwards. Liang Sicheng, having experienced criticism, anti-criticism and self-criticism, passed away in 1972 when the Cultural Revolution was not even over. While Kenzo Tange, as technique bureaucrat architect, was entrusted with important post in some state projects such as Olympics in Tokyo and Osaka World Expo and finished assignment.

Both of Liang Sicheng and Kenzo Tange had studied international architecture designs, and investigated and analyzed their own national traditional heritages respectively. Their ideologies on politics became an inevitable problem for they had built buildings that can be considered as state symbols. However, the People's Heroes Monument and Hiroshima Peace Monument have maintained one representing under huge pressure and complex situation by imitating stone tablet and Haniwa respectively. In my opinion, the reason is that both of them had beliefs on carrying-out technicism architecture of the idea of "art=architecture=city=state=people". Judging from the present which by no means matches this equation from the perspective of profession or ideology, only "a man's courage" that bursts from such historic moments as revolution of new China or city doom caused by A-bomb can support the beliefs. After that Liang Sicheng passed away in the criticism of culture revolution; while Kenzo Tange left Japan and became imperial architect of a Middle East country. Our society does not need the past bureaucrat architects any more.

建筑评论的向度

李翔宁 邓圆也

> 除了要懂点儿建筑，其他几乎一切东西我都还必须懂点儿。
> ——路易斯·芒福德 (Lewis Mumford)

引言

题记出自 1975 年芒福德（Lewis Mumford）写给费舍尔（Thomas Fisher）的回信。在信中他谈到自己如何成为一名建筑评论家。[1] 在建筑评论界长期对现代性及其衍生观念的讨论中，历史、艺术、地理、社会科学等内容的高度理论化影响着建筑评论的方式和结论，建筑评论吸收着跨学科的知识。这种思潮不断演进的同时，观念的边界也开始模糊和交错，对人类居住环境的更立体的思考进入了建筑评论的视野。如今，我们处在一个既疏离又杂糅的时代，不是芒福德的高度文化黏性的时代，也非赫克斯特布尔（Ada Louis Huxtable）写作初期的专业媒体时代。我们所处的时代前所未有地激进、庞杂、充满流动性；我们承担着全球化带来的一切后果，而当下的地方性经验恰恰充满了冲突和博弈。当我们发现以西方为中心建构起来的话语体系不能描述和解释我们身边的诸多建筑和文化现象，当对柯布西耶和路易·康的研究无法为中国传统建筑研究或当代乡土问题提供参照，我们开始关注世界上和我们有类似境遇的其他地区：在西亚、非洲、南美，前殖民时代的传统和后殖民话语的双重参照，让我们有机会在全球化的现代性话语体系之外，建构一个特殊性、地方性的知识系统。

2015 年 12 月 11 日—14 日，在同济大学建筑与城市规划学院和中国美术学院建筑艺术学院召

Dimensions of Architectural Criticism

Li Xiangning, Deng Yuanye
Translation: Wu Yan (from Chinese to English)

> In addition to knowing something about architecture,
> I needed to know something about almost everything else.
> —Lewis Mumford

Introduction

This inscription is from a letter written by Lewis Mumford to Thomas Fisher in 1975, in which he speaks of his path to becoming an architectural critic. When modernity and its derivative concepts are discussed in the field of architectural criticism, highly theoretical disciplines such as history, art, geography, and the social sciences influence its methods and conclusions, for architectural criticism draws on a pool of interdisciplinary knowledge. At the same time as this ideological trend has continued to develop, distinctions between different concepts have started to intersect and blur. A more dynamic concern with human living environments has entered the lexicon of architectural criticism. Today, we find ourselves in an age of alienation and hybridity, no longer the culturally-cohesive era of Mumford, nor the era of professional media during which Ada Loise Huxtable first launched her writing career. We live in an age of unprecedented amounts of radicalism, disorder, and mobility; we bear the full burden of globalization, while local experience is rife with conflict and negotiation. Once we came to the realization that Western-centered discourse had failed to articulate and interpret the architectural and cultural phenomena surrounding us, once the studies of Le Corbusier and Louis Kahn were

开的"2015国际建筑评论家委员会(CICA)研讨会"[1]（下文简称"CICA研讨会"），以"超越东西南北——当代建筑中的普遍性与特殊性"为主题，邀请了来自北美、南美、欧洲、亚洲、大洋洲、非洲的建筑评论家、学者以及建筑师，尝试通过分享世界不同文化背景的案例来观察当代建筑的普遍性与特殊性。研讨会探讨了当代建筑评论的全球向度，亦为中国当代建筑应对文化转型这一议题提供了启发和教益。会议也对社会公众开放，希望进一步通过建筑评论加强建筑学的公共性和社会性。

建筑具有时间性，它是动态的。这种时间性不仅体现在历史背景上，也体现在建造和使用的过程中。不同文化背景所包裹的不同宗教观、文化价值、政治体系和自然生态特征，与全球化进程共同作用，对城市建筑产生了有趣的"在地全球化"的影响。在宗教文化比较强大的社会中，信仰往往与自然环境息息相关，而宗教信仰又与文化政治导向相关联，这样的特点在伊斯兰地区尤为显著。对于现代资本主义国家而言，建筑与自然环境、历史沿革、社区自发和赋权的关系受到广泛关注。在大多数政治权力集中、社会治理以自上而下的权力运行为主导的地区，意识形态和政治对文化乃至实践的建构影响之深远亦不容忽视。

通过比较研究，笔者尝试印证建筑与自然生态共生、与意识形态并进的动态模式，而CICA研讨会通过各大文化带的实例基本论证了这种趋势，对于中国当下的建筑批评乃至建筑实践本身，都具有借鉴意义。

地方性知识：当代建筑中的普遍性与特殊性

在这次会议上，研讨嘉宾分享了不同的案例研究，展现出丰富的研究视角。除了传统的欧洲城市之外，来自拉美、非洲、亚洲的案例都令人耳目一新。这些地区的建筑对现代主义都自有一套消化系统，同时，地区差异在建筑上的投射清晰而丰富。

曼努埃尔·夸德拉（Manuel Cuadra）通过极富文学性的语言描述了城市建筑背后隐藏的人

---

[1] 国际建筑评论家委员会（Comité International des Critiques d'Architecture，简称"CICA"）集合了全球各地的建筑评论家、建筑理论家和建筑史学家，是国际上重要的建筑评论家组织之一。"建筑评论和理论话语是构成建筑进程整体的必要部分"，秉持这种理念的CICA评论家们致力于以评论和理论服务建筑，并与建筑师一起服务城市。CICA有多样化的渠道将话语与城市建筑结合，包括与政府和非政府组织合作。委员会认为，更理性的城市规划、建筑和景观设计能够提升人居生活质量，而实现这一目标需要在整个设计过程中纳入大量的批判性讨论和理论话语。委员会与媒体合作也十分密切，尤其是与规划、建造和建筑类的杂志学刊，委员会旨在打破建筑话语在公共舆论界仅作为建筑作品新闻纪实的局限，以此促进建筑话语的专业性、理论性和公共性。除此之外，委员会还强调建筑评论需要探讨建筑的社会面向和文化面向，亦不可绕开建筑在历史中的角色。

found unable to provide reference for research into traditional Chinese architecture or contemporary rural issues, we began to shift our focus to other regions of the world that share a situation similar to our own: West Asia, Africa and South America. Pre-colonial traditions and post-colonial discourse provide dual references, allowing us to establish a specific, local knowledge system free from the dominance of the global discourse of modernity.

On December 11-14, 2015, the College of Architecture and Urban Planning at Tongji University and the Department of Architecture at the China Academy of Art hosted the "International Committee of Architectural Critics (CICA) Shanghai 2015 Symposium"[1]. With a theme of "Beyond Orient and Occident, North and South: The Universal and the Particular in Contemporary Architecture", architectural critics, scholars, and architects from North America, South America, Europe, Asia, Oceania, and Africa were invited to the symposium as a way to survey the notion of universality and particularity in contemporary architecture by sharing case studies on diverse cultural backgrounds from all around the world. The symposium shed light on the global dimension of contemporary architectural criticism and offered contemporary Chinese architecture inspiration and lessons in dealing with the issue of cultural transformation. Furthermore, the symposium was open to the general public, and hoped to use architectural criticism to heighten the public and social aspects of architectural studies.

Architecture is temporal and dynamic. This temporal character manifests itself not only in historical context, but also in the process of construction and employment. Diverse cultural backgrounds, which include differing religious views, cultural values, political agendas, and natural ecological features, act in concert with the progress of globalization, exerting an interesting "glocalization" influence on urban architecture. In societies where religious culture plays a relatively dominant role, beliefs are often closely tied to the natural environment, and religious beliefs are intertwined with cultural-political orientation, most notably in Islamic regions. Modern capitalistic countries, meanwhile, are widely concerned with architecture's relationship to the natural environment, the course of history, community initiatives, and empowerment. In regions that tend towards totalitarianism and top-down social manage-

---

[1] Comité International des Critiques d'Architecture (CICA) brings together critics, theorists and historians from all over the world. The members of CICA share the conviction that criticism and the theoretical discourse of architecture in general are integral parts of the architectural process. Consequently, CICA and its members see themselves as working towards a better architecture and better cities alongside the architects.
CICA offers cooperation with governmental and non-governmental institutions working towards raising the quality of life through improved urban planning, architecture and landscape design. The aim is to include critical debate and theoretical discourse as a part of the design process.
CICA offers cooperation with the press, and particularly with the magazines dedicated to planning, building and architecture. The goal here is to promote an architectural discourse that goes beyond journalism and the documentation of buildings, and emphasizes the social and cultural dimensions of architecture and its role in a historic perspective.

与自然的关系以及由此发展出的社区精神，从早期人类将自然景观转化为文化景观的劳作场景到欧洲中世纪城市社区的形成，乃至文艺复兴时期大型城市规划的出现；从殖民时期，南美洲本土文化被限制而体现出欧洲化倾向，到巴西城市建筑发展中对"食人主义"的借用……夸德拉生动而具体地描绘了地景、城市、建筑与人类生活之间相互缔造的关系。

露斯·沃得·采恩（Ruth Verde Zein）介绍了2014年"罗杰里奥·萨尔莫纳拉美建筑奖：开放空间/集体空间"的评选过程和获奖作品——FGMF建筑事务所设计的圣保罗Projeto Viver住宅项目。采恩认为，这座建筑体现了该奖项"建筑应该塑造城市与公共空间"的评奖宗旨，而塑造公共空间是解决拉丁美洲复杂的城市问题的最佳途径。

阿克塞尔·索瓦（Axel Sowa）选择了三个位于巴黎且与建筑和城市空间遗产的重新利用有关的案例：拉维莱特圆形画廊餐厅（The Rotunda of La Villette）、拉维莱特公园（La Villette Park）和Musée précaire Albinet临时博物馆。它们的再利用过程体现了建筑和城市规划可以作为调解空间、政治、人的活动之间关系的工具。

森古·奥依门·古尔（Sengül Öymen Gür）以土耳其博德鲁姆的德米尔别墅群为例，深入阐释了伊斯兰地区现代建筑与自然文化的共生关系。建筑师图肯特·坎瑟沃（Turgut Cansever）将伊斯兰宗教观和哲学观中的时间和空间关系转移到建筑中成为一种"场所精神"。与德米尔别墅群遥相对望的安缦如雅度假酒店，由土耳其建筑师艾米·奥贡（Emine Ogun）和默罕默德·奥贡（Mehmet Ogun）设计。霍里·伊布拉海·杜曾利（Halil Ibrahim Duzenli）对这个项目进行了分析，认为该建筑同样表达了伊斯兰文化逻辑，并较好地处理了人的行进、建筑空间和自然环境之间的关系。

凯伦·艾克尔（Karen Eicker）以南非约翰内斯堡的一个项目为例，就当代建筑的普遍性与特殊性展开讨论。该项目位于一个高密度人口聚居区——霍尔布罗地区，它建造于20世纪70年代，是当地第一个社会基础设施项目。该项目试图与社区和城市肌理相呼应，看上去是一个很"轻"的建筑。

李翔宁关注当代中国独立建筑师的实践，展示了几十位建筑师的代表作品，包括张永和、王澍、刘家琨等引领的"实验建筑"思潮之后的青年建筑师的实践。李翔宁分析了这些建筑师的职业特征，其中独立性、年轻化、先锋、边缘、当代性和如何表达"中国性"（Chineseness）是他们的主要特点。这些建筑师的建筑理念从实验性跨越到批判性实用主义，且仍在不断变化。李翔宁描绘了一幅动态的中国独立建筑师实践的全景图画。

金秋野通过对建筑师庄慎、王澍的几个案例的分析，提出对于中国当代建筑的"自然几何学"（Natural Geometry）思考。他指出，这种自然几何学虽然具有现代主义的思想，但在融入了中国的来世观念、日常观念、古典审美之后，在建筑表达上被柔化了。自然几何学将中

ment, ideological and political implications for both culture and its practical development are far too significant to be ignored.

Through comparative research methods, the authors of this article have attempted to confirm the dynamic modes of both architecture's cohabitation with natural ecology and co-development with ideology. Indeed, using case studies taken from diverse cultural territories, the CICA symposium put forth strong arguments for the existence of such trends, providing an invaluable source of reference for current Chinese architectural criticism and even architectural practice itself.

Local Knowledge: The Universal and the Particular in Contemporary Architecture

Over the course of the symposium, guest speakers presented a wide range of case studies, reflecting an ample variety of research angles. In addition to classic European cities, cases from Latin American, African, and Asian cities provided the audience with a brand-new perspective. Architecture from each of these regions has developed its own set of assimilative apparatus for modernism. At the same time, local characteristics can be seen clearly and in great diversity on such architecture.

In highly literary terms, Manuel Cuadra described the relation of man to nature that is concealed behind urban architecture, and the community spirit that evolved from it, starting from the scene of early humans transforming natural landscapes into cultural landscapes, touching on the emergence of urban communities in the cities of Antiquity and the Middle Ages, the establishment of large-scale urban planning during the Renaissance, and the colonial period, when limiting the expression of local traditions in South America reflected the influence of the European way of life, to the appropriation of "cannibalism" in the development of urban architecture in Brazil. Cuadra delivered an expressive and specific account of the reciprocal relationship between landscape, city, architecture, and human life.

Ruth Verde Zein introduced the selection process for the 2014 "Rogelio Salmona Latin American Architecture Award: Open Spaces / Collective Spaces" as well as the winning project – the Projeto Viver Building in São Paulo, which was designed by FGMF Architects. Zein believed this project was representative of the guiding mandate of the award – that architecture

国哲学思维中的自然观溶解在了日常实体之中，并以建筑的语言描绘出来，且仍保持其人文性。

王骏阳讨论了现代中国语境下的建筑理论与实践。通过一些"奇奇怪怪"的建筑案例，王骏阳看到了部分中国当代建筑对于自身价值观的投射，以及对于现代性的想象。他敏锐地提出了转型时期建筑理论的责任和面临的问题，例如如何应对转型时期建筑理论的动荡和理论的失落和复归；认为建筑理论是建筑思想积累的过程，在这个过程中，如何建立多元开放的建筑理论体系是理论界应该关注的问题。王骏阳还阐述了对理论在实践中的应用的思考。

卢永毅以20世纪二三十年代的上海为窗口，考察早期西方现代建筑进入上海后的传播和本土化发展，超越将"现代建筑作为某种统一的文化模式的认知的局限"，重新认识这段西方现代建筑在近代中国移植与转化、实践与想象的历史过程，并从中提炼现代性经验。

周榕分析了中国当代本土建筑实践中的三个亭子——直向建筑的"采摘亭"、大舍建筑设计事务所的"花草亭"和冯纪忠的"何陋轩"，通过分析三者表现出的不同的文化观念、设计和建造方式等内容，对"建构"（Tectonics）思想在中国的研究进行了反思。

刘晨则从艺术史和图像研究的角度，将两个艺术作品做了两种比喻：一是通过分析影像作品中一个抽象的人物形象，揭示当代中国社会文化的西方化和自我文化身份认同的缺失；二是通过一个金字塔的装置隐喻西方博物馆制度和它传达的西方模式对当代博物馆这一公共空间和文化形式的侵蚀。

葛明关注当代中国建筑教育，尤其是设计教育中对于现代性的话语转化，以及对西方建筑教学模式的吸收与消化，并通过自己的建筑教学案例对以上问题进行深入剖析。

回响：当代建筑评论的向度

更迭的城市理性

城市从现代性走向"过度现代性"（over modernity）的首要特征就是其城市理性的不断更迭。

16世纪起，城市理性的基础之一是宗教。符合宗教教义的建筑旨在通过几何关系来验证上帝与凡人的关系。该建构隐喻在现代建筑中仍有余热，如身为加尔文派教徒的柯布西耶所作

should help create city and public spaces – and argued that shaping public spaces can be the best approach to resolving complex urban issues in Latin America.

Axel Sowa chose three case studies from Paris related to the adaptive reuse of architectural and urban-space heritage: the Rotunda of La Villette, La Villette Park, and the Musée Précaire Albinet. The process in which these spaces were reused reflects how architecture and urban planning can play a role in mediating the relationship between space, politics, and human activities.

Using the example of the Demir Houses in Bodrum, Turkey, Sengül Öymen Gür offered his insights into the cohabitation relationship between modern architecture and natural culture in the Islamic world. Architect Turgut Cansever transforms the sense of time and space found in Islamic religious and philosophical views into a "spirit of the place". Across from the Demir Houses is the Amanruya Resort, designed by Turkish architects Emine Ogun and Mehmet Ogun. Halil Ibrahim Duzenli used an analysis of this project to argue that this design similarly expresses Islamic cultural logic and artfully manipulates the relationship between human movement, architectural space and the natural environment.

Karen Eicker used a project in Johannesburg, South Africa to open up discussion on the universality and particularity of contemporary architecture. Built in the 1970s, the project is situated in Hillbrow with high population density and is considered the first local social infrastructure project. Designed to resonate with the character of the local community and the urban tissue of the area, the project demonstrates a "light" architectural form.

Li Xiangning reflected on the work of contemporary Chinese independent architects and showcased dozens of representative works, including the accomplishments of young architects that followed the "experimental architecture" movement led by Yung Ho Chang, Wang Shu, and Liu Jiakun. Li analyzed the professional characteristics of these architects, showing how their identity was shaped by independence, youth, avant-gardism, marginality, and contemporaneity, as well as the question of how to express "Chineseness". Leaping from experimentalism to critical pragmatism, the design concepts of these architects have continued to evolve. Li illustrated a dynamic panorama of independent Chinese architects in practice.

By examining examples from architects Zhuang Shen and Wang Shu, Jin Qiuye proposed the notion of "Natural Geometry" reflected in contemporary Chinese architecture. He pointed out that, while natural geometry involves modernist ideas, the inclusion of Chinese notions of the afterlife, everyday life, and classical aesthetics have neutralized its architectural expres-

的《直角之诗》（Poème de l'Angle Droit）和科斯塔（Lucio Costa）为巴西利亚所作的充满神秘主义色彩的规划。又如德米尔别墅群和安缦如雅酒店的空间关系投射了伊斯兰的宗教观和世界观，并通过物体关系制造了空间，又通过控制物体与人的关系制造了场所。

CICA研讨会特邀嘉宾矶崎新谈到了梁思成与丹下健三如何建构国族性和将国家意识的顺从深刻地投射到建筑设计当中。综观具有民族主义传统的国度，其城市理性的动机基本建立在"皇权"限定了"疆土"的客观事实之上。城市是意识形态和政治诉求所描绘的功能性图景。政权的合法性和制度的公正性以"神圣的空旷"（holy emptiness）来表达，规训功能借由对几何关系的控制结构体现出来。

基于上述两类城市理性所建造的理想几何模型，时刻以实体与流动的社会现实对击。实体地理环境、伦理审美需求以及社会政治经济现实，这三种因素始终存在着辩证互动的关系。社会形态之于个体议题、客观之于主观、权威力量之于社区家庭，这三组关系也存在于城市实体之中。例如夸德拉的巴西"食人主义"建筑的后殖民议题以及卢永毅对近代上海的研究；又如索瓦分析了法国的政治社会事件对建筑影响，以及建筑对公共生活的调解媒介角色等。这些案例均阐明了三组关系在城市建筑实体之中的博弈。

演进的主体意义

建筑评论一直都在寻找主体意义。20世纪的评论学者关注新城市生活中的实用主义。例如，芒福德尝试连接专业知识与大众认知，让城市和建筑变得可读。之后，贝尔德（George Baird）对平等主义的有效性提出质疑，认为虽然信息传达方式平等，但其接收后不一定能被平等地消化。与平等主义启蒙者芒福德相较，赫克斯特布尔则认为评论家必须居间公正地存在于建筑师—艺术家和使用者—观者之间，更关注项目的细节、建成和意义。此时，平等主义理念明显缓和了，史学性和地方性知识快速增长。

随着建筑评论的学科发展，一些学者例如保罗·戈德伯格（Paul Goldberger）认为建筑评论既是美学的，也是政治的、社会的和文化的，它们紧密相连作为理解建筑的书写基础。而复杂合作和共同责任的产物——建筑，能比任何一种艺术形式更好地解释人类行为，帮助我们寻找当代问题的本质。CICA研讨会的众多案例，就评析了实践作品存在于既定事实和随机观察之外的部分，即所谓"不言自明"与"不言不明"的部分。

多向度的当代性

今天，我们不得不迎来"全球化"。一些学者认为全球化会抹杀身份和文化特征，导致"去地域主义"。兰格（Alexandra Lange）认为全球化使学科不断膨胀而缺乏深度，缺乏能阅读建筑的历史、情绪、肌理和建造的评论。然而，查克·克娄斯（Chuck Close）反倒认为全球化和

sion. Natural geometry dissolves Chinese philosophical views on nature into the substantiality of everyday life, while further expressing them in the language of architecture, yet sustains its sense of humanism.

Wang Junyang discussed architectural theories and their practice in the context of modern China. By introducing examples of "weird" architecture, Wang Junyang perceived some contemporary Chinese architecture as being a projection of its own sense of value and imagination of modernity. He keenly identified the responsibilities and problems faced by architectural theory during transition periods, such as how, during such periods, to deal with the turbulent state of architectural theory and its subsequent losses and recovery – he proposed that architectural theory is the process of accumulation of architectural beliefs, and that during this process the theorists should concern themselves with the question of how to develop a dynamic, open system of architectural theory. Wang further expounded on the application of theory in practice.

Lu Yongyi used Shanghai in the 1920s and 30s as a window to study the proliferation and local development of early Western modern architecture when it first entered Shanghai. Going beyond the "limited perception of modern architecture as some kind of unified model of culture", he revisited the history of modern Western architecture in modern China, and its transposition and transformation, practice and imagination during this process, as a way to extract out experiences of modernity.

Zhou Rong reviewed three pavilion projects from contemporary Chinese local architectural practice – the "Harvest Pavilion" by Vector Architects, the "Blossom Pavilion" by Atelier Deshaus, and the "Helou Pavilion" by Feng Jizhong. By analyzing the differences in their cultural conception, design, and construction, Zhou's presentation reflected on "tectonics" research in China.

Starting from the perspective of art history and iconography, Liu Chen made two metaphors about two different works: first, by analysing an abstract human figure in one video work, he exposed the Westernization and absence of cultural identity in Chinese culture; second, by suggesting a pyramid installation as a metaphor for Western museum institutions and the Western methods they promote, he criticized how the contemporary museum as a public space and a cultural form has been corrupted.

Ge Ming's presentation was concerned with architectural education in China, especially pedagogical approaches to the changing discourse of modernity and the adaptation and assimila-

超媒体化会激发学者去寻找更多的操作空间。

笔者认为,全球化并不是抹杀特殊性的原因,它是政治经济技术整体发展的结果,并不天然地具有全球性。早在全球化之前,人类早已具备了营造普遍性的能力。而特殊性在全球大同的幻觉之中时常会更为突出。一方面,族群重建自我认同使得身份边界更加清晰;另一方面,在意识和物质都杂糅的时代,杂糅本身成了文化生产和输出的重要力量——不仅是生产模式和技术的混合,更是不同利益链和文化价值的协调。正如吉安卡洛·德卡罗(Giancarlo De Carlo)所说,"建筑太重要了,不能只留给建筑师。"

在当代城市中,场所主义的、服务于沟通和公共性的空间不断被营建起来。从空间上看似乎打破了阶级壁垒,填满了"神圣的空旷";然而从城市肌理上看,"城市地质分层"仍然存在,差异只是暂时地在集体性空间中面目模糊,实际上它们并没有隐去,其在具体生活中仍旧清晰。所以,通过这些现代主义主流视野之外的案例,我们尝试在具体之中寻找对人的境况更贴近的解释系统。资本主义和威权主义在普遍意义上不断地影响着人们实际生活的模样;而建筑,虽是人们生活实体化的一个重要载体,它却也无法背负城市化进程的一切责任。作为建筑师和建筑评论家,亦无法只面对建筑设计这一个向度。所以,建筑评论要保存建筑内涵的多向度,并在历史的爬行中,向学者和公众展示其中不可见的事实(unseen truth)。

tion of Western models of architectural education. Using examples from his own teaching experience, he offered an in-depth analysis of these questions.

Resonance: Dimensions of Contemporary Architectural Criticism

Shifting Urban Reason

Cities' movement from modernity towards "over-modernity" is primarily characterized by constantly shifting urban reason.

Since the 16th Century, religion has stood as one basis of urban reason. Architecture that conforms to religious doctrine must verify the relationship between gods and mortals through geometry. This kind of tectonic metaphor still can be seen in modern architecture, for example, in how Corbusier – a Calvinist – built the Poème de l'Angle Droit, or how the influence of mysticism permeates Lucio Costa's plan for Brasilia. Or in the case of the Demir Houses and the Amanruya Resort, where spatial relations reflect Islamic religious outlooks and worldviews. Space was produced out of the relationship between objects; place by controlling the relationship between objects and humans.

At the special request of the CICA symposium, Arata Isozaki spoke of how Liang Sicheng and Kenzo Tange constructed notions of nationalism and projected the compliance of nationalist sensitivity into their architectural designs. Overall, in countries with nationalist traditions, the impetus behind urban reason is essentially founded on the objective fact that "imperial power" defines "territory". Cities are functional prospects depicted by ideological and political aspirations. Legitimacy of the regime and impartiality of the institution are manifested in the form of "holy emptiness", and disciplinary function is embodied in the control over geometric relationships.

The ideal geometrical models constructed by the two types of urban reason mentioned above are constantly in physical contact with the flux of social reality. A dialectic, reciprocal relationship between physical geography, ethical and aesthetic needs, and the social, political and economic reality has always existed. The relationships between social form and individual subject, objective and subjective, and authoritative power and community and family can also be found in the physical body of a city. For example, it can be found in the post-colonial themes of Cuadra's "cannibalism" architecture in Brazil, in Lu Yongyi's research on modern

参考文献Reference:

[1] Fisher, T. The death and life of great architecture criticism[J]. Places Journal, 2011(12): 3.

[2] Segre R., Scarpaci L., Coyula M.. Havana. Two Faces of the Antillean Metropolis[M]. North Carolina: The University of North Carolina Press, 2002: 49.

[3] Davis, A. Identity crisis: estrangement in the evolution of architectural criticism[M. Wolfgang F. E. Preiser, Aaron T. Davis, Ashraf M. Salama, Andrea Hardy. Architecture Beyond Criticism: Expert Judgment and Performance Evaluation. London: Routledge, 2014.

[4] [法]米歇尔·福柯. 规训与惩罚：监狱的诞生[M]. 刘北成, 杨远婴译. 北京：生活·读书·新知三联书店, 2003：86.

[5] Diderot, D. Salon of 1765 and Notes on Painting[M]. New Haven: Yale University Press, 1765: 12-24.

[6] Fisher, T. The death and life of great architecture criticism[J]. Places Journal, 2011(12): 2.

[7] Baird, G. Thoughts on the current state of criticism in architecture[J]. Journal of Architectural Education, 2009, 62(3): 5.

[8] Goldberger, P. Why Architecture Matters[M]. New Haven: Yale University Press, 2011: 171.

[9] Huxtable, A. On Architecture: Collected Reflections on a Century of Change[M]. New York: Walker, 2008: 69.

[10] Lange, Alexandra. Why Nicolai Ouroussoff is not Good Enough[J]. The Design Observer, 2010(2): 3.

[11] Giancarlo De Carlo. Architecture's public[J]. Parametro 5, 1970: 4-12.

[12] Attoe, W. Architecture and Critical Imagination[M]. Chichester: John Wiley, 1978.

[13] Benedikt, M. On the Role of Architectural Criticism Today[J]. Journal of Architectural Education, 2009, 62(3): 6-7.

[14] Derrida, J. White Mythology: Metaphor in the Text of Philosophy[M]. Derrida J. Margins of Philosophy. Chicago: University of Chicago Press, 1982: 220-253.

[15] Fisher, T. Making Criticism More Critical[J]. Journal of Architectural Education, 2009, 62(3): 12-13.

[16] Goldberger, P. Architecture Criticism: Does it Matter?[Z]. Paul Goldberger, 2015.

[17] Heynen, H. Architecture and Modernity: A Critique. Cambridge, Massachusetts: MIT Press, 1999.

[18] Lange, A. Writing about Architecture[M]. New York: Princeton Architectural Press, 2012.

[19] Levinson, N. Critical beats[J]. Places Journal, 2010(3).

[20] Marcinkoski, C. and J. Arpa. The City That Never Was[Z]. The Architecture League of New York, 2012.

[21] Moscavici, C. Diderot's salons: Art Criticism of Greuze, Chardin, Boucher and Fragonard[M. Moscavici, C. Romanticism and Post-romanticism (excerpt), 2010.

[22] Mumford, L. The Culture of Cities. Boston: Mariner Books, 1938.

[23] Mumford, L. Architecture as a Home for Man: Essays for Architectural Record[M]. J.M. Davern, ed. New York: Architectural Record Book, 1975.

[24] Robert Wojtowicz, ed. Sidewalk Critic: Lewis Mumford's Writing on New York[M]. New York: Princeton Architectural Press, 1998.

[25] Ouroussoff, N. Koolhaas. Delirious in Beijing[N]. New York Times, 2013-07.

[26] Parsons, G. Fact and Function in Architectural Criticism[J]. The Journal of Aesthetics and Art Criticism, 2011, 69(1): 21-29.

[27] Sorkin, M. Critical Mass: Why Architectural Criticism Matters[J]. The Architectural Review. 2015(5): 28.

[28] Sontag, S. Against Interpretation[M. Sontag, S. Against Interpretation and Other Essays. New York: Dell, 2009: 13-23.

[29] 常青. 建筑学的人类学视野[J]. 建筑师, 2008（6）：95-101.

[30] 李翔宁, 张晓春. 浅议城市生态规划及其在中国的发展方向[J]. 现代城市研究, 1999（2）：11-13, 21-63.

[31] 李翔宁. 当代城市的五副面孔——城市转型国际论坛与城市理论的进展[J]. 建筑学报, 2007（8）：42-43.

[32] 李翔宁. 论道：对当代中国建筑的再思考[J]. 建筑学报, 2014（3）：90.

[33] 卢永毅. 建筑：地域主义与身份认同的历史景观[J]. 同济大学学报, 2008（1）：39-48.

[34] 卢永毅. 遗产价值的多样性及其当代保护实践的批判性思考[J]. 同济大学学报（社会科学版）, 2009（5）：35-43.

[35] 帕塔比·拉曼, 理查德·柯因, 王凯, 郑时

Shanghai, and in Sowa's account of how political and social events have influenced architecture and how architecture is a mediator for public life. Each case study expounds on the negotiations between these three distinct relationships within the physical body of urban architecture.

Evolution of Subjective Meaning

Architectural criticism never stops pursuing subjective meaning. Scholars of 20th century concern themselves with pragmatism in new urban life. For instance, Mumford explored the possibility of connecting professional knowledge with public perception, such that cities and architecture could become readable. Later, George Baird questioned the effectiveness of egalitarianism, suggesting that egalitarian principles, when communicated, are not necessarily egalitarian upon reception. In contrast to Mumford's egalitarian enlightenment, Ada Louise Huxtable asserted that critics should find their position somewhere between architect – artist and user – viewer and pay more attention to detail, construction, and meaning. By then, development of the notion of egalitarianism had obviously slowed, and significant improvements can be noticed in the development of historical and local knowledge.

Accompanying the disciplinary development of architectural criticism, scholars such as Paul Goldberger argue that architectural criticism is aesthetic as well as political, social, and cultural. All of these are intertwined and constitute the basis of writing as a way to understand architecture. A product of intricate collaboration and common responsibility, architecture is more capable of explaining human behaviour than any other form of art, helping us locate the essence of contemporary issues. Indeed, the case studies brought up at the CICA symposium investigate parts of architectural practice that exist outside of established facts and random observations – the so-called "self-evident" and "self-unknown" parts.

Multifold Dimensions of Contemporaneity

Today, we have no choice but to welcome "globalization". Some scholars assert that globalization will obliterate identity and cultural characteristics, resulting in "de-regionalism". Alexandra Lange believes that globalization will cause academic disciplines to expand in breadth but lack in depth, losing the criticism that connects us to a building's references, emotions and textures. But Chuck Close argues that globalization and hypermediatization might stimulate scholars to seek out additional room in which to operate.

This paper's authors think that globalization is not to blame for the obliteration of particu-

龄.建筑批评的生产[J].建筑师,2006(3):38-46.

[36] 唐克扬.当代中国建筑的第三条路[J].建筑学报,2014(3):100-102.

[37] 郑时龄.当代中国建筑的基本状况思考[J].建筑学报,2014(3):96-98.

[38] 郑时龄.理论与建筑批评——关于建筑批评的读书笔记[J].建筑与文化,2012(2):8-13.

[39] 王骏阳.建筑实践与理论反思[J].建筑学报,2014(3):98-99.

[40] 王骏阳.建构的话语、建造的诗学与建筑史学[J].时代建筑,2012(2):37-41.

[41] 周榕.具足的批判性建筑——主导知识再组织的当代建筑批评图式[J].世界建筑,2014(8):32-37.

larity. Rather this is a result of joint political, economic, and technological development, which is not naturally global. Long before globalization, humans already possessed the ability to generate universality. Particularity appears more prominent when seen in the illusion of global homogeneity. On the one hand, ethnic groups' rebuilding of their self-identity has resulted in more clearly defined identities; on the other hand, in an age where mind and matter have started to blend, this blending itself has become an important force of cultural production and output – not only the mixture of production methods and technology, but also the coordination of different interests and cultural values. As Giancarlo De Carlo put it, "Architecture is too important to leave to the architects."

In contemporary cities, spaces that are place-oriented, serve communication needs, and public in nature are constantly being built. Spatially speaking, they seem to have broken down class barriers and filled in the "holy emptiness". However, when one considers the urban fabric, "urban stratification" is still present, just temporarily disguised in the form of collective spaces. This stratification is far from faded, and still distinguishable in the specifics of day-to-day lives. Thus, through case studies arising from outside the realm of mainstream modernist viewpoints, we try to identify ways of interpretation that derive from specific situations and are grounded in human conditions. Capitalism and authoritarianism, as types of universality, are continuously shaping the real lives of people; but architecture, although an important medium for the materialization of people's lives, cannot bear the burdens of urbanization alone. Architects and architectural critics, too, can no longer focus one-dimensionally on architectural design. Therefore, architectural criticism should promote the multifold dimensions of architectural meaning, revealing unseen truths to academics and the public alike as history passes by.

认同 —— 人之境况的建筑：
由五个篇章和一个尾声构成的思考

[秘] 曼努埃尔·夸德拉

译者：莫万莉（英译中）

校译：吴彦

空间、物质和时间

《2001 太空漫游》（*2001: A Space Odyssey*）是一部 20 世纪 60 年代末期由斯坦利·库布里克（Stanley Kubrick）[1] 导演的电影，它反思了当今世界中技术与经济的高速发展和社会与文化的激烈冲突并存的重要历史时期。第一幕"人类的黎明"的宇宙背景暗示着宇宙的起源。从大爆炸的那一刻开始，时间、空间与物质成形了，也正是这三个类别构成了建筑可操作的范畴。伴随着理查德·施特劳斯（Richard Strauss）[2]《查拉图斯特拉如是说》（*Also sprach Zarathustra*）的音乐，一系列静态图片一张张地出现，来传达这层含义。在这些图片中，我们看到的是初生的地球，一个动植物尚未出现的世界。除了构成自然界的基本矿物质外，它一无所有，但我们知道正是这些矿物质构成了我们生活的世界和人类生命体。

在长达数百万年的物种进化过程中，植物和动物依次出现，最后出现的是哺乳类动物和灵长类动物。灵长类动物有和人类关系最为接近的祖先，与我们有许多相似之处。在接下来的场景中，我们看到了一个奇异的、完全不同的地方。我们发现了一块仿佛来自另一个世界的、具有完美几何形状的黑色巨石。它在一群灵长类动物中产生了强烈的魔力。与此同时，库布里克为我们展现了一个太阳、月亮与地球排成一线的宇宙现象——日蚀。突然间，类人猿发生了巨大的变化，它们开始变得暴力，用自己创造的工具互相残杀。这一场景与那块充满魔力并具有完美几何形状的神秘巨石一同出现，而这块巨石在本质上几乎就是建筑。

接着，一只类人猿将一枚骨制武器抛向空中来宣告自己的领土，随即这根骨头变幻成为了一

---

[1] 斯坦利·库布里克（1928—1999），美国著名导演——译者注。
[2] 理查德·施特劳斯（1864—1949），德国杰出浪漫派作曲家及指挥家——译者注。

Identity – The Architecture of the Human Condition: A Reflection in Five Movements and One Epilogue

Manuel Cuadra
Translated by Line Umbach (from Spanish to English)
Edited by Mary Anne

Space, Matter, and Time

Stanley Kubrick's *2001: A Space Odyssey* from the late 1960s reflects an important period in the history of the contemporary world: a time of fast technological and economic development as well as of intense social and cultural conflicts. The universal context of the opening scene, entitled "The Dawn of Man", alludes to the origin of the universe – the "Big Bang" from which time, space and matter came into being. Architecture also operates in these three categories. The accompanying music, Richard Strauss' "Also sprach Zarathustra" parallels with a series of static images which is drawn out, in order to communicate this message. In these images we are confronted by the Earth at its beginning, a world still free of flora and of fauna, consisting of nothing but the basic mineral essence of nature. We understand that this mineral matter is the substance of the world we live in and of human life itself.

Over the course of millions of years, evolution brings plants and animals, and finally mammals and primates – the immediate predecessors of man which bear similarities to us. In the next scene, we are confronted with a strange and completely different place. We see an otherworldly perfectly geometrical black monolith which has a huge fascination to a group of primates. At the same moment, Kubrick confronts us with a cosmic phenomenon. The sun, moon and earth align in a solar eclipse. Suddenly a profound transformation takes place among the apes as they become violent and begin to kill each other with the tools they have created. This scene is juxtaposed with the fascinating and mysterious perfect geometric monolith which is almost architectural in nature.

艘宇宙飞船，场景突然跳跃了几千年的时间来到了一个虚构的现在。在这个世界中，人类与外太空的互动已经成为常态。几个宇航员在山丘上观察一座放射状的月球城市克拉维斯基地（Clavius Base），而地球则在远处的地平线外。这一场景展示了人类在月球上永久定居的景象，而从外太空观察地球已经成为一个寻常的现象。

从类人猿向空中抛掷骨头到月球上的场景，经历了几千年剧烈的文化演变。现在我们将从城市层面上来观察这一文化演变的过程。

空间、物质和时间
Space, matter and time

The scene suddenly jumps thousands of years ahead to a fictitious present time: a humanoid being throws a weapon made out of bone into the air in order to claim a territory for himself. In this moment the bone is transformed into a space ship. In this world, man's interaction with outer space has become normal. A permanent presence of man on the moon is depicted by a scene in which some astronauts observe from a hill a radial lunar city "Clavius Base", and the earth appears beyond the horizon in the distance. Viewing the earth from the outside universe has become common place.

Between the humanoid throwing a bone into the air and the scene on the moon, thousands of years of intense cultural development has passed. We will now look at this process of cultural development at the level of the city.

Vernacular Urbanism – The Constitution of the Urban Community in the European Middle Ages

The architecture of the early city was influenced by community and the political structure of society. Buildings were built by people, house by house, with little direction or imposition from an outside power. In this sense, the constitution of the community preceded the physical constitution of the city.

In the year 1547, Vitruvius' "De Architectura libri decem" was translated into French by Jean Martin. In this edition the tractate includes some drawings by Jean Goujon, an artist who translated his theories about the origins of architecture into drawings. The first of these drawings depicts early man in two contrasting scenes. In the background a group of naked men and women, perhaps the descendants of Adam and Eve expelled from paradise, are walking through a lush forest. We observe how the presence of man has carved a new linear pathway through the forest, and how this turns a natural landscape into a cultural landscape, which includes clear signs of human presence. And we note how Goujon presents this scene with a detailed picturesque landscape with mountains and a valley in the far distance.

The people in the foreground of the drawing (clearly centuries later in time) have stopped their tour and rest in a group around a fire. Some are wearing incipient clothes. The elimination of the trees has created a circular space where they have settled to cook on the fire in

地方城市主义——欧洲中世纪时期城市社区的构成

早期城市的建筑受到社区和社会政治结构的影响，房屋往往由人们一座座地亲手建造，很少受到外部力量的指示或是摆布。从这种意义上来说，社区的形成要先于城市的物质性形成。

1547年，让·马丁（Jean Martin）将维特鲁威的《建筑十书》翻译为法文版。这一版本包含了一系列由艺术家让·古戎（Jean Goujon）绘制的插图，以图像的形式呈现出维特鲁威关于建筑起源的理论。第一张图描绘了早期人类的两个对比鲜明的场景。在背景中，一群裸体的男人和女人在一片繁茂的树林中行走，他们或许是被逐出天堂的亚当和夏娃的后人。我们可以观察到人类的出现是如何在树林中开辟出一条笔直的新道路，以及它是如何将自然景观转化成为文化景观的，人类存在的痕迹被清晰地呈现出来。我们也注意到古戎是如何通过对远方群山与峡谷的优美景色的细致入微的描绘来展示这一场景的。

前景中的人物则显然来自几个世纪之后。他们中止了行进，正一起围着篝火休憩，其中一些人穿着原始的服装。树木被砍倒，形成了一块圆形的空地，使他们能够安顿下来并在空地中央的篝火上烹饪。古戎的画表现了通道（或道路）和城市广场的起源。正是人类的行动、行走和休憩定义了最早的永久聚落，而道路和城市广场则随后成为这些永久聚落的构成要素。

接下来的两幅插画展示了人们通过实践来学习如何建造防御屏障和有遮蔽的住所。这些"人造洞穴"由木材和泥土这些能够在现场找到的自然材料搭建而成。因此，房屋与它周围的自然景观具有一种直接的关系。在图中，我们可以看到人们通过对几何学的运用来提高建造技能并发展合理的建造体系。

背景中繁茂的树木形成了前景中小型聚落的陪衬，凸显了自然的存在。这正是村落发展的过程：树木被砍伐用作建筑材料，而自然景观和大地的地形特征却被保留了下来。人类通过利用特定地点的天然优势来与大自然和谐相处，比方说他们会在山顶的位置欣赏风景。

让·古戎的画在这一时期戛然而止，没有描绘接下来的第一批城市建立和发展的阶段。

而我们则将目光转向公元前5世纪的雅典。雅典城的规划包括了三个基本元素：首先是结构复杂的居住区，它包含了顺应地形曲折蜿蜒的街道和由其形成的不规则街区；其次是雅典卫城或者说上城，经过理性的布置与设计的高耸神庙形成了一座众神之城；最后是位于城市最下方的阿格拉（Agora），它是市集和日常公共生活的中心。

有趣的是，这和让·古戎的插画有着相似之处。在两种情境中，周边景观和地形都起到了十分重要的作用。在雅典，早期居民有意识地选择了地形中的制高点来建造堡垒，并围绕它形成聚落。在这之后，由于城市社区的成长、扩张以及民主化的政治发展，堡垒演变为守护城

the middle of the space. Goujon reflects the origins of the path (or road) and of the town square which become the later constituent elements of permanent settlements originally defined by doing, walking, and resting.

A second and a third drawing show the people, learning by doing, how to build shields and covered shelters. These "artificial caves" are made out of wood and mud – the natural resources found at the site. Therefore, the relationship of the buildings to the surrounding natural landscape is very direct. We see the people developing building skills and rational building systems by making use of geometry.

There is also a very strong presence of nature depicted as lush trees in the background which form a backdrop to the small settlement in front. This is how villages would have developed. Trees are cleared and used as building materials and yet the natural landscape and the topography of the land remains. Man works in harmony with nature by taking advantage of the positive aspects of the natural local – for example enjoying the views from the location on top of a hill.

Jean Goujon stops at this point. He does not represent the next steps, which would have showed the founding and development of the first cities.

Instead we look to the city of Athens in the 5th century B.C. The plan of Athens consists of three basic elements. First, the residential area with its complex structure consisting of irregularly meandering streets which follow the topography of the land and create irregular blocks. Second, the Acropolis or upper city, which consists of its very rationally placed and designed towering temples and forms a city for the gods. Third, the Agora in the lowest part of the city is the marketplace and the centre of everyday public life.

Interestingly, there are parallels between this representation and those by Jean Goujon. In both the surrounding landscape plays an important role, as does the topography. In Athens, the early people consciously chose this place by forming their settlement around the highest point of the landscape to build a fortress. Later, due to the growth expansion of the community and the political development towards democracy, the fortress became the city for the gods who watched over the city. A new peaceful society based on a common faith, common rituals and religious traditions replaced the oppression and violence of the early settlers. The establishment of a direct visual connection between the Acropolis and the Agora expresses a very deliberate positioning of these places in relation to each other in the landscape.

SIENA

复杂性和矛盾性
Complexity and contradiction

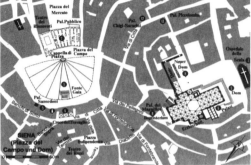

市的众神之城。一个崭新的、建立在共同的信仰、仪式和宗教传统之上的社会的平和状态取代了早期定居者之间的压迫与暴力。卫城和阿格拉之间直接视觉联系的建立，体现了对二者在地形中相对位置关系的精心考虑。

这些结构和房屋承载着社区的精神。城市的每一个部分和相应的房屋都有其特定的大小、规模与秩序。居住区规模较小，有着柔和的有机秩序，而公共区域和神圣的神庙则有着更大的规模和精确的几何秩序。

雅典是欧洲地中海地区早期城市发展的一个典范。罗马帝国覆没后长达几个世纪的混乱之后，在中世纪，欧洲城市的建立过程不得不重新开始。位于欧洲南部的锡耶纳正是这样的一个案例。它不规则的城市结构是当时欧洲城市社区的共同特征。这些城市社区中的人们享有根据自己的生活方式选择居住方式的自由。

最初，锡耶纳由分别围绕三座山丘形成的三个社区组成。随着时间的推移，这三个社区连成了一片。当人们开始在山谷中建造城市，公共空间便出现了。作为刚刚形成的社区公共空间，位于山谷的坎波广场（Piazza del Campo）成为了一个举行诸如赛马这样的城市庆典的场所。在那时，赛马被认为是一种能够和平解决城市社会中的争端的活动。坎波广场体现了社区生活和由其产生的城市空间以及城市建筑之间的紧密关系。在锡耶纳，市政厅所象征的公共生活中心和大教堂所象征的宗教生活中心之间也具有强烈的视觉联系。

场所、地形和其他环境之间的关系以及人们如何在这一环境中组织和表现他们的生活方式，在这些方面锡耶纳与雅典有着明显的相似之处。通过不规则的住宅街区、狭窄的街道和这些街道与公共空间之间的关系，城市的规划体现出城市的复杂本质。这也是欧洲中部城市的特征，它们中的大多数建立于公元9世纪至11世纪。这些城市的中心城区所具有的这一典型特征也影响到了我们今天构建城市的方法。城市结构不仅仅意味着一种实体呈现，它们的大小、规模、秩序和物质性都传达着一种建立在参与、共同信仰和共同制度之上的社区精神。

学院城市主义——南美洲对城市和文艺复兴式建筑的被迫接受

在南美洲，我们面对的则是一种完全不同的情况。南美洲没有经历过欧洲意义上的"中世纪"，西班牙在南美洲开辟殖民地之前，它曾有过初步的城市发展，但随即被西班牙征服者完全地压制了。因此，墨西哥和秘鲁原有的城市结构被从欧洲引进的城市主义所取代。从这种意义上来说，南美洲被迫接受了欧洲城市以及文艺复兴和巴洛克式的建筑。因而从它们

These structures and buildings support the spirit of community. Each part of the city and the corresponding buildings have their particular size, scale, and order. The residential area has a small scale and a soft organic order, and the public areas and holy temples have a larger scale and a precise geometric order.

Athens is an example of an early city development in Mediterranean Europe. After the chaotic centuries following to the fall of the Roman Empire, in the Middle Ages, in Europe the process of the foundation of cities had to start again. The city of Siena is an example of this in southern Europe. The irregular urban structure is characteristic of European communities of that time in general. These communities enjoyed the freedom of defining the way they occupied their territory in accordance with their way of life.

In the beginning, Siena consisted of three hills, each with its own community. Over time, the three communities grew together. As they built in the valley a common space developed. In the valley the Piazza del Campo, a public space of the newly constituted community, became a place for celebrating urban rituals such as the Palio, thought to peacefully solve conflicts inside the urban society. This shows a strong relationship between community life and the resulting urban space and thus the architecture of the city. Again there is a strong visual relationship between the centre of public life, as represented by the Palazzo Pubblico, and religious life, as represented by the Cathedral.

The similarities to Athens are evident in the connection between the place, with the topography and other environmental conditions, and the people organising and expressing their way of life in community in this context. The plan of the city expresses its complex nature through the irregular housing blocks, the narrow streets and the connections of these streets to the public spaces. This is the kind of urbanism that characterizes the cities in central Europe too, most of which were founded in the 9th, 10th and 11th century. Most of the central areas of all these cities typically have this character which has influenced the way we structure our cities today. Urban structures communicate more than a physical presence. Their size, scale, order and materiality all communicate the spirit of the community – a spirit which is based on participation, a common faith, and common institutions.

的基本形式上看，南美洲城市并非随着城市社区的建立而来，而是被迫接受了文艺复兴和巴洛克的城市模型而形成的结果。

人文主义是文艺复兴理想的一个关键。在莱昂纳多·达·芬奇（Leonardo da Vinci）的《维特鲁威人》一画中，他研究了人体的本质和支配它的规律。他将这些关系简化为一种理想的几何语言。与此相似，这一时期的城市和建筑抛弃了中世纪城市的有机复杂性而朝着理想的几何语言发展。

作为建筑师，我们往往会偏好几何形式、规律性、网格和柏拉图立体，但我们不能忘了人文主义仅仅是文艺复兴的一个侧面，而它的另一个侧面则是重商主义。与中世纪通过农业来产生财富的典型方式相比，重商主义通过货币经济和贸易构成了一种抽象的产生财富的方式。这导致了一个新的精英阶级的形成，这些富有而有权势的家族具有影响一个城市的建造的能力。他们雇佣建筑师来规划和设计"他们的"城市。由此，具有完美几何形式的"理想城市"诞生了，这种城市理念与中世纪时期背道而驰。理想城市的设计象征着一种人为中央集权的出现，而正是这种集权主宰着民众，并试图去控制他们的生活和环境。尽管完美几何平面的和谐之感依然存在，但它和环境的自然地形之间的矛盾也变得显而易见。自然总是以复杂的形态来表达自身，而这恰恰与几何形式的"精确性"特征相矛盾。

本文将这种形成城市的方式称为"被迫接受的城市"（the imposition of the city）。我们可以观察到这种预先设定的城市结构剥夺了人们原有的选择在何处居住以及如何搭建住所的自由。相反地，人们不得不接受忽视个人偏好而分配给他们的住所。它不再是一座由城市社区所形成的城市，而是一座由中央集权所主宰的城市。无论在整体层面还是个体层面上，社区都被迫接受了中央集权强加于其的生活方式。

相比欧洲，这种城市在南美洲殖民地产生了更为剧烈的影响。以古巴的哈瓦那为例，在被西班牙人殖民后，它成为了从秘鲁和墨西哥产出的稀有金属矿石被运输至西班牙过程中的一个重要的收集和转运点。随着非洲奴隶被输送到古巴，它渐渐地开始为欧洲生产更多的商品。古巴人民的生活总是被欧洲权力中心的利益所左右。因而从这层意义上来说，古巴人民的生活方式是被迫接受的。

这对于哈瓦那的城市和建筑肌理有着直接的影响。哈瓦那的早期规划显示了它的第一批建筑是一些要塞和军事建筑。它们的规划呈几何形态，由石头砌筑而成，周边则是居住区。在较早的时候，这些居住区中曾出现过一定的不规则的形态。但随后文艺复兴式规则的正交网格被附加于这些区域。在类似教堂或是皇宫这样一些象征中央集权的建筑上，我们也可以发现精确的几何语言和石材的运用，以及对那一时期欧洲式风格和模式的采纳。

西班牙殖民统治于1898年结束，此后古巴陷入了和美国的、具有强烈新殖民式色彩的依附

Academic urbanism – The Imposition of the City and Renaissance Architecture in America

In America we are confronted with a completely different situation. There were no Middle Ages in the European sense there. Before Spanish colonisation there were original developments, that later were completely suppressed by the Spanish conquerors. As a consequence, European urbanism was brought to America and substituted the urban structures in Mexico and Peru. In this sense, there was an "imposition of the city" and of Renaissance and Baroque architecture in America. Thus American cities in their basic form were not preceded by the constitution of a community. They were the result of the imposition of Renaissance and Baroque urban and models.

One key Renaissance ideal is Humanism. Leonardo da Vinci, in his drawing the "Vitruvian Man", studies the nature of the human body and the laws which govern it. He reduces these relationships to the ideal language of geometry. In a similar way, the urbanism and architecture that emerged from this period of time moved away from the organic complexity of the medieval city, towards the ideal language of geometry.

As architects we tend to be preoccupied by geometric forms, regularity, grids, and Platonic solids. But we must remember that Humanism is only one facet of the Renaissance. Another facet is Mercantilism. Compared to the Middle Ages' typical production of wealth through agriculture, Mercantilism constitutes an abstract way of generating wealth through a money economy and trade. This formed a new set of elites – rich and powerful families who had the power to influence the construction of a city. They commissioned architects to plan and design "their" cities. The result was the opposite conception of the city compared to Middle Ages. The concept of a perfectly geometric "Ideal City" came into being. The design of the Ideal City expresses the presence of a central man-made power that dominates people and pretends to control their life as much as the landscape. A sense of harmony persists, yet conflicts come to light between the perfect geometrical plan and the natural topography of the landscape, for example. Nature always expresses itself in complex forms and this conflicts with the "exact" characteristics of geometric forms.

This way of generating cities is called here "the imposition of the city". We observe that this kind of predefined urban structure eliminates the original liberty of people; the freedom to decide where to live and how to build their own houses. Instead, people are obliged to use the exact places assigned to them, independent from their personal preferences. It is not the city developed by the urban community. It is the city dominated by central power which im-

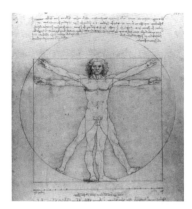

合理性和被迫接受
Rationality and imposition

关系中。这时的城市建筑开始反映最新引进的城市理想和模型。作为一个孤立的体块，哈瓦那的卡皮托利乌国会大厦脱离于城市系统。从形式上说，它很容易让人联想到华盛顿特区的国会大厦。

在20世纪50年代后期，我们可以观察到一种"美国化的生活方式"和"美国梦"的影响。娱乐建筑反映出一个社会对通过消费获得享乐的需求。出于对这种需求的回应，大量电影院、俱乐部和旅馆被建造起来。位于维达多区的高耸的哈瓦那希尔顿酒店和它周边的其他建筑即是例证。这些奢华的建筑掩饰着城市其他区域贫民窟的贫穷和极端恶劣的生活状况。在1959年的政治革命后，古巴最终脱离美国获得了独立。

心灵与身体的融合——"食人主义"建筑在巴西

与古巴不同，巴西曾是葡萄牙的殖民地，但它也经历了类似的转变。在面对新殖民文化依赖之时，巴西哲学家、巴西现代主义的创始者之一奥斯瓦尔多德·安德拉德（Oswaldo de Andrade）提出了所谓的"食人主义运动"（Anthropophagic Movement）。他把"文化食人主义"（cultral cannibalism）的概念描述为一种将"他者"的文化融合进来的方式。这种"他者"的文化既包括巴西国内和国外的盎格鲁—美利坚文化，也包括来自其他大洲的移民和美洲原住民的文化。与巴西精英阶层所倡导的对外来模式和生产方式的依赖相反，他提出对各种外来文化产物和技术进行侵略性的挪用和内化。他认为在此基础上才有实现一条既能为自身所用又能对外输出的本土文化生产链的机会。

巴西的海岸、雨林、引人入胜的地形、动植物以及南美洲原住民的丰富文化形成了它令人惊叹的自然之美。虽然巴西建筑受到了葡萄牙的巨大影响，然而鲜明的环境使其殖民风格建筑形成了独有的特征。以里约热内卢为例，它有着非常分明的地质景观和十分特别的城市文化，正是这一极具震撼力而又独特的环境启发了安德拉德。对于他来说，为这样一种环境想象和创造一种新的文化、艺术和建筑时，丝毫没有复制过去风格的必要。

奥斯卡·尼迈耶（Oscar Niemeyer）也拒绝复制过去的风格。位于里约热内卢的教育与健康部大楼是他的第一个重要项目。它建成于20世纪30年代末期，源自尼迈耶与勒·柯布西耶以及景观建筑师罗伯托·布雷·马克思（Roberto Burle Marx）的合作。尼迈耶之后的作品，即1940年代在潘普利亚（Pampulha）和20世纪五六十年代在巴西利亚（Brasilia）的作品，清晰地显示出受到奥斯瓦尔多德·安德拉德启发的"食人主义"设计策略的影响。他在潘普利亚的圣弗朗西斯阿西西小教堂（Saint Francis of Assisi chapel）中采用了混凝土壳体。然而

poses a way of living on the community as a whole and on individuals.

This influence played out more dramatically in the colonies than in Europe. Havana in Cuba, for example, which was colonised by the Spanish, became a central point of collection for the precious metals mined from Peru and Mexico, which were then shipped from Havana to Spain. Later, slaves were brought from Africa to Cuba and gradually Cuba began to produce more goods for Europe. The life in Cuba was always defined by the interests of the centres of power in Europe. In this sense it was an imposed way of life.

This had a direct influence on the urban and architectural fabric of Havana. An early plan of the city reveals that the first buildings were fortressed and military buildings, geometrically planned and built in stone. These were surrounded by the housing area which at an early point, had certain irregularities. Later a regular orthogonal Renaissance grid was imposed over this area. In the architecture of the churches and the palaces of the representatives of the central power, we observe precise geometry and a realization in stone, too, and the application of the European styles and models of the time.

After the end of the Spanish colonial period in 1898, Cuba was marked by a strong neo-colonial dependence from the United States. Again the architecture of the city began to express the new imported ideals and urban models. In Havana, the "Capitolio" – the site of the parliament, sits as an object in isolation and does not form a part of a larger urban system. Formally, it is particularly reminiscent of the Capitol building in Washington DC.

Later in the 1950s, we observe the influence of the "American way of life" and the "American dream". Recreational buildings reflected the desires of a society that consumes for enjoyment. In response to this desire, cinemas, clubs and hotels were built. The tall Havana Hilton Hotel and other buildings in its surrounding Vedado district are examples. These luxury buildings belie the poverty and extremely bad living conditions of the slum areas in other parts of the city. In 1959, Cuba gained independence from the United States through political revolution.

心灵和身体的融合
Intergrating mind and body

Integrating Mind and Body – The Architecture of Anthropophagy: Brazil

In contrast to Cuba, Brazil was a Portuguese colony, but the country went through a similar transformation. In response to neo-colonial cultural dependence, Oswaldo de Andrade, a philosopher and one of the founders of Brazilian Modernism, created the so-called "Anthropophagic Movement". He described the concept of "cultural cannibalism" as a way to integrate the culture of the "others", the Anglo-Americans both inside and outside the country, as well as immigrants from other continents or Native Americans. Instead of relying on imported models and production, as did the Brazilian elite, he proposed an aggressive appropriation and internalization of all kinds of imported cultural goods and techniques. On that basis, he saw the opportunity to develop a local and cultural production chain for own use and export.

Brazil has a striking natural beauty – beaches, rain forest, interesting topography, flora and fauna, as well as the rich cultures of the Native Americans. There is a strong Portuguese influence in the architecture, but in the strong Brazilian context, the colonial buildings take on their own unique and individual character. Rio de Janeiro, for example, has a very definite physical landscape, and a very particular urban culture. It makes sense that Oswald de Andrade was inspired by this powerful and individual context, and found no sense in copying the styles of the past when imagining and creating a new culture, a new art and architecture.

Similarly, Oscar Niemeyer refused to copy the styles of the past. His first big work, the Ministry of Education and Health in Rio de Janeiro from the late 1930s, resulted from a cooperation with Le Corbusier and the landscape architect Roberto Burle Marx. Later works from the 1940s in Pampulha and the 1950s and 1960s in Brasilia, clearly show his "anthropophagic" approach to design, as inspired by Oswaldo de Andrade. When designing the small Saint Francis of Assisi chapel in Pampulha, he used a concrete shell. But he designed it with a plasticity and ornamental character reminiscent of the colonial vaults, as opposed to the highly technical finish of modern shells. From a rationalist and constructivist European point of view, he could be considered "just" a formalist, but this did not personally concern Niemeyer. His reinterpretation of Mies van der Rohe's Barcelona Pavilion in his own house in Las Canoas, built in the 1950s in the middle of the forest on a rock and with a view of the Atlantic, is like an architectural translation of German to Brazilian: it preserves the same spirit while passing from an orthogonal geometry to organic forms.

The new Capital City of Brasilia which he designed together with Lucio Costa, was inspired

与现代壳体采用高超工艺的完成面不同，他的设计让人联想起殖民时期穹顶的造型和装饰特征。从欧洲式的理性主义和建构主义观点来看，他或许"仅仅"能被视作一名形式主义者。但尼迈耶并不关心这一点。他位于拉斯堪诺阿斯（Las Canoas）的自宅，可以说是对密斯·凡·德·罗设计的巴塞罗那馆的重新诠释。这座建于1950年代的屋子位于树林之中的一块巨石之上，俯瞰着大西洋。它如同一次从德国式到巴西式的建筑转译，在将正交几何转化为有机形式的过程中保留了同样的精髓。

巴西的新首都巴西利亚由尼迈耶与卢西奥·科斯塔（Lucio Costa）共同设计。它在受到凡尔赛规划的影响的同时，也受到了墨西哥的前西班牙城市主义的影响。最为重要的是，巴西利亚融入到了巴西中部独特的自然景观之中，每一幢建筑被置于其中却并不打破它的连续性。这是一座极度延展的汽车城市，与1950年代的规划思想相符合。然而正如1789年之后的凡尔赛，这一理念也很快在现代城市中被质疑和摒弃。巴西利亚巨大的城市开放空间折射出了它们所处的广阔疆域的尺度，从而形成可使用的公共中心。

在建筑细节上，尼迈耶创造出了一种受到古希腊和罗马主题（由此是欧洲式的）直接影响的形式语言，并在此之中融合了一些现代理念。在由总统办公大楼、国会大厦和最高法院形成的三权广场中，巴西式的柱式和低矮的三层楼高的幕墙表现出了一种不同的思考方式。对于一个理性的、来自欧洲的观者来说，或许这会产生一种奇特的超现实的并置关系。这样一种建筑有着截然不同的逻辑：这是一种"食人主义"的逻辑。正因为如此，巴西利亚是一个独树一帜的创造，它成功地转译了它所诞生的那个时代的独特精神。

挪用的建筑——发现体系之外的南美"中世纪"

巴西城市肌理的另外一面则由贫民区构成。奇怪的是，里约热内卢的贫民区却以其所占据的风景优美之地而闻名。它们或是位于山丘之上，或是位于树林之中，往往有着望向周边城市和大海的极佳视野。它们毗邻最为豪华的住宅区，贫民区的居民们则为后者提供各种服务。正如让·古戎所描绘的早期聚落，贫民区也由从周边环境获得的材料建成，只不过这些材料是城市建造的废弃材料罢了。尽管如此，贫民区是居民们一栋房子接着一栋亲手建造的，这也包括它的街道和公共空间。它们形成了贫民区的城市结构，并使之成为一个整体。由此，它与中世纪城市结构之间的相似之处显而易见。在这两个例子中，城市结构都反映出了社区的生活方式。在贫民区中，这种生活方式与巴西社会的一些严重的经济和社会问题相关。然而，城市由居民建造而成的理念对于发展中国家来说是具有借鉴价值的。

by Versailles, as much as by the Mexican pre-Hispanic urbanism. Most importantly the city is integrated into the unique natural landscape of central Brazil: the single pieces of architecture were placed without interrupting the continuity of the territory. The idea of an extremely extended car-city corresponds to the thinking of the 1950s, which soon lost its validity and practicality in the modern city – much like Versailles after 1789. The large open urban spaces are scaled to reflect the size of the enormous territories they serve, thus creating usable public centres.

In the architectural detail, Niemeyer creates a formal language very directly inspired by ancient Greek and Roman (therefore European) themes, incorporating modern ideas. The Square of the Three Powers with the Palácio do Planalto (presidential office), the Congresso Nacional (national congress), and the Supremo Tribunal Federal (supreme court), with its columns in a "Brazilian order" in combination with three-storey low curtain walls, all reflect a different way of thinking, sometimes generating – for the rationalistic European eyes – strange surrealistic constellations. The logic behind such an architecture is a different one: it is an anthropophagous logic. As such Brasilia is an original creation and a successful translation of the unique spirit of the time of its origins.

The Architecture of Appropriation – Discovering the American "Middle Ages" beyond the System

The other face of Brazil's urban fabric is given by the favelas. The favelas in Rio de Janeiro, strangely enough, are well known for the beautiful sites they occupy, on the hills, in the forest, and with incredible views of the surrounding city and the ocean. They sit right next to the finest residential areas, where the favelados offer their services. Like the early settlements shown by Jean Goujon, the favelas were built with the materials from the direct surroundings, only in the latter case the materials are the construction waste of the city. Nevertheless, the favelas are cities built by the people themselves, house by house, including the streets and public spaces which create the urban structure for the favela as a whole. So the similarities of the favela to the medieval structures are evident. In both cases, the structures express the way of life of the community. In the case of the favelas, this way of life is related to some severe economic and social problems of Brazilian society. However, the idea of a city made by people is a valid model for developing countries.

体系之外的建筑
The architecture beyond the system

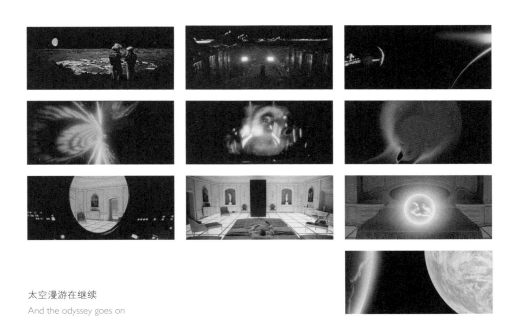

太空漫游在继续
And the odyssey goes on

...and the Odyssey Goes on — An Epilogue

But Stanley Kubrick's "Odyssey" goes on. After the scenes showing the "Ideal City", scientists discover the cuboid (from the first scenes of the movie), during their geological explorations and excavations of the Moon surface. And they discover a pathway between this cuboid and a place far away in the Universe. So they organize a tour to this place in order to find out what is there. The following scenes are dedicated to the internal life of the space ship. We are then confronted with flashing images, light, colours and forms — apparently this is Stanley Kubrick's way of translating what it could mean to travel at a speed close to the speed of light into film images. Then we see the main protagonist of this tour expressing fear, and we follow his view point as he looks through the window of the space ship first into the surrounding outer space.

But suddenly we there outside see a room with a surprising design that combines modern elements such as an illuminated glass floor with European historical ornaments, Rococo paintings and furniture. The protagonist is part of this scene: he is lying in a bed, looking extremely old and seems close to dying.

And he directs his eyes to the cuboid which now stands in front of the bed, and appears strong and sharp with its orthogonal geometry. Suddenly the orthogonal geometry appears everywhere. It appears in the orthogonality of the grid of the blanket and of the whole bed, the orthogonality of the grid of the illuminated floor and of the whole chamber, and in the orthogonality of the furniture and of the paintings. Everything emanates the geometric essence of the cuboid — contemporary elements as much as those associated with European historical art and culture.

Finally the protagonist appears instead of as an old man, as a foetus. First inside a bubble on the bed, then inside a bubble floating in the space with the foetus watching the Earth from the outside.

In ending, we are left wondering if this was really a physical journey through the Universe — or if this has been a journey to the most inner parts of ourselves. We are left wondering whether there may be no difference between the material and the spiritual dimensions. Whatever we do as humans has a material dimension as it has an inner dimension related to our spiritual life, to the inner life as we experience it, each person for themselves.

太空漫游在继续——一个尾声

斯坦利·库布里克的"太空漫游"仍在继续。在"理想城市"的场景之后，科学家们在对月球表面的地质探索和挖掘过程中，发现了第一个场景中出现过的长方体巨石。他们发现了这块巨石和一个位于宇宙深处的地点之间的通道。为了探索这一地点，他们组织了前往那里的远征。接下来的场景呈现太空飞船内部的生活。我们看到一系列闪现的画面、光线、色彩和形状。显然，这是斯坦利·库布里克用电影影像的方式来诠释当人们以接近光速运动之时会发生的景象。接下来，我们看到了这次远征的主人公显露出恐惧的神情。随着他的视线，我们第一次透过太空飞船的舷窗看到了周围的外太空。

突然间，我们到了一间设计奇特的房间。它结合了诸如发光玻璃地板这样的现代元素和欧洲历史上的装饰元素，洛可可风格的绘画和家具。主人公也成为了这一场景的一部分。他躺在床上，看上去极其苍老，似乎濒临死亡。

他望向立于他床前的长方体巨石，其正交几何外形看上去坚固而锋利。突然间，正交几何无处不在。它出现在了毯子的网格图案和整张床的垂直线条之中，出现在了发光地板的网格和整个卧室的垂直线条之中，出现在了家具和绘画的垂直线条之中。无论是当代元素或是与欧洲历史上的艺术与文化相关的元素，所有的一切都源起于长方体巨石的几何本质。

最终，主人公与其说看上去像一位老人，不如说更像是一个胎儿。他处于一个床上的气泡之中，也处于一个漂浮于太空的气泡之中。在这里，他从外部望向地球。

在电影结束之时，我们开始疑惑这是否是一段真实的穿越宇宙的旅行，抑或仅仅是发生在我们内心最深处的旅行。我们开始质疑物质维度和精神维度之间的区别。作为人类，无论我们做什么，正如我们的行为总有一种物质维度，也总有一种与我们的精神世界、与我们每个人所体验的独特的内心世界相关的内在维度。

（标*图片来源：Jean Goujon，Vitruvius；

标**图片©Manuel Cuadra；

标***图片来源：Segre Havanna；

标****图片来源：UIA Habana 1963；

标*****图片来源：Rio Do Cosmografo ao Satelite；

其他图片由作者提供。）

(Photo copyrights:

*: Jean Goujon，Vitruvius；

**: ©Manuel Cuadra；

***: Segre Havanna；

****: UIA Habana 1963；

*****: Rio Do Cosmografo ao Satelite；

The rest of the photos are provided by the author.)

建筑及其调解的力量

[德]阿克塞尔·索瓦
译者：陈迪佳（英译中）
校译：邓圆也

在《古迹》（Antiquity）一书中，普遍性和特殊性的哲学问题被描述成一种不平衡的关系：一个不可分割且不可改变的整体如何才能包含某一类单体呢？如何才能确定这种普遍的整体无形地存在着呢？普遍性是仅存在于我们的意识中，还是我们可以在现实世界中遇到它呢？[1]

建筑批评家极少能上升到这样的高度，在如此澄澈明净的纯抽象哲学领域中从事工作。他们需要用自己的方式将普遍性的问题与具体的生活环境相联系。于是，建筑批评不得不同时在抽象概念和物质形式的领域中展开。批评只能在无所不在的共同价值和实实在在的建成作品之间小心翼翼地行进。

如果我们试图以这种方式重新建构普遍性的问题，那么便会涉及社会行为中关于规范性的方面。普遍有效性的积极作用可以在众多协议中找到，例如1972年签署的关于保护世界文化和自然遗产的公约，或者1999年国际建筑师协会（UIA）关于建筑实践专业水准的条约。然而，无论这些被普遍认同的协议多么具有约束力，它们都不足以控制建筑和城市设计领域中所有可能出现的情况。从多种可能的行动中做出选择，正是建筑批评家在谈论和写作的内容。评论家通常只应付那些既定经济和政治条件限制下以及特定空间环境下的特殊的设计和规划难题。最重要的是，建筑批评家关注自己的判断。只有不被搅入特定案例的特殊性中，他们的判断才能超越武断而吹毛求疵的学究迂腐。当我们考虑到那些可以作为相似案例的一以贯之的评判基础的原则时，具有一般性（generality）的"普遍性"就可以发挥作用了。

本文试图剖析一些近年来法国建筑和城市规划中的实践和理论。笔者选择的案例位于巴黎

---

[1] 关于"普遍性"概念的总体介绍详见：http://plato.stanford.edu/entries/universals-medieval/#spade85.

Architecture and the Power of Mediation

Axel Sowa

The philosophical problem of the universal and the particular occurs in *Antiquity* with the description of an unbalanced relation: How can a class of singular things be contained in an indivisible and supposedly immutable entity? How do we gain certainty about the intangible existence of such universal entities? Are universals only in our minds or can we encounter them in the physical world? [1]

Architectural critics can hardly embark on such altitude flights through cloudless philosophical spheres of pure abstraction. They need to reformulate the problem of universals in their own way by relating it to particular living environments. Architectural critiques necessarily operate in relational fields exposed to both, abstract concepts and material forms. Criticism can only be pertinent as a vigilant navigation between the ubiquitousness of shared values and the concreteness of built forms.

If we try to reformulate the problem of universals in this sense, normative aspects of social action come into sight. Values of universal validity can be found in agreements such as the convention on the protection of world cultural and natural heritage, signed in 1972, or the 1999 UIA accord on professionalism in architectural practice. However binding these universal agreements are, they do not suffice to govern all possible actions in the fields of architecture and city planning. Choices made out of a multitude of possible actions are precisely what architectural critics speak and write about. Critics tackle particular design and planning decisions, which are made under given economic and political constraints, and within singular

[1] For a general introduction to the notion of universals see: http://plato.stanford.edu/entries/universals-medieval/#spade85 (accessed on March 4th, 2016)

拉维莱特圆形画廊餐厅
Rotounda la Villette

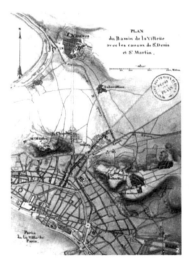

左　拉维莱特圆形画廊餐厅与圣马丁运河 平面
上中　拉维莱特圆形画廊餐厅与圣马丁运河
上右　拉维莱特圆形画廊餐厅与圣马丁运河 总平面
下中　拉维莱特圆形画廊餐厅
下右　拉维莱特圆形画廊餐厅与地铁二号线
LEFT　Plan of Rotounda la Villette & Canal St. Martinv
UPPER MIDDLE　Rotounda la Villette & Canal St. Martin
UPPER RIGHT　General plan of Rotounda la Villette & Canal St. Martin
LOWER MIDDLE　Rotounda la Villette
LOWER RIGHT　Rotounda la Villette & Metro Line 2

spatial situations. Above all, architectural critics are concerned with their own judgements. If their judgement is supposed to reach beyond the arbitrary pedantry of fault-finders, it should not be absorbed by the particularities of specific cases. The generality of universals come into play when we consider principles that could serve as a base for consistent judgements of similar cases.

In this paper, I will provide insights into recent theories and practices of French architecture and urban planning. The cases I selected for this purpose are all situated in the northeastern part of Paris and its adjacent suburbs. Within this urban zone, social and economical transformations were most drastically felt over the last decades. They affected a territory, anciently occupied by food and manufacturing industries. Economic change led to perpetual renegotiations of functional boundaries and social divisions, conveyed by a post-industrial urban condition, which France shares with many other European countries. First signs of this condition, such as industrial disinvestment, increase of unemployment, and individualization of life styles had been recorded since the mid-1970s, after a long period of economic growth. If we look at French societal debates of the last decades, we can identify topics that are largely and repeatedly discussed. Among those we find notions of cultural heritage, leisure industry and participation, which exemplify different ways to cope with the post-industrial condition. Since the early 1980s, collective memories, the right to individual pleasure, and increased chances of political partaking had been also discussed in architecture and urban planning. If we look at architecture not only as a spatial framework of daily life, but also an embodiment of aesthetic and political thoughts of its time, we may ask how these social and political issues affect the operational modes, by which our environment is built or transformed.

Heritage

In France, cultural heritage enjoys great attention and growing prestige. In 1984, Jack Lang, former French minister of culture, inaugurated the heritage days, allowing French citizens and foreign visitors, to inspect the nation's architectural treasures. The success was immediate and tremendous. In 1991, European Union claimed patronage over the idea, which had been copied by many other states, since. Alone in France, every year, more than 15 million people come to visit what they collectively own. The annual economic turnover of French heritage industry, including maintenance, repair and tourism is estimated at more than 20 billion Eu-

的东北部和附近的郊区。过去几十年间,这一片区的社会经济变革最为剧烈,影响了这个自古以来以食品业和制造业为主的地区。透过法国及其他许多欧洲国家的后工业城市状况,可以看到经济变革导致了对功能边界和社会群体划分的不间断的重新协商与磨合。其征兆从20世纪70年代中期,经历了长时间的经济增长后,开始显现出来,例如工业撤资、失业增加以及生活方式的个体化。我们可以从过去几十年里法国社会的争论中辨识出广泛的、被普遍讨论的一些话题。这其中包括关于文化遗产、娱乐产业和公众参与的主张,它们代表了应对后工业状况的不同方法。从20世纪80年代初开始,集体主义、个体享乐的权力及参与政治的机会的增加也成了建筑和城市规划谈论的问题。如果我们不仅仅将建筑看作日常生活的一个空间框架,还将其看作这一时代的美学和政治思想的载体,那么我们也许会好奇,这些政治和社会事件是如何影响那些建设和改变我们周遭环境的运作模式的。

遗产

在法国,文化遗产享有广泛的关注和越来越高的的声誉。1984年,法国前文化部长杰克·朗(Jack Lang)设立遗产日,让法国民众和外国游客得以一窥这个国家的建筑瑰宝。这一举动获得了巨大的成功。1991年,这个想法得到了欧盟的资助,而后被其他国家效仿。仅在法国,每年就有超过1500万人前来参观他们共有的财产。法国遗产产业每年的资金周转约超过200亿欧元,其中包括维护、修复和旅游业。[2] 回顾这条道路,它的成功是在差异(discrepancy)中铸就的。在经历了法国大革命的创伤,许多城堡、教堂和艺术作品遭到暴动分子的破坏之后,这个年轻的共和国对危在旦夕的遗产进行了清点盘查。在暴乱中幸存下来的建筑物以公益的名义被分类、测量和修复,而后被宣布为法国国家财产。[3] "建筑遗产"在法语中被称为"patrimoine",这个词源于"patrimonium",古罗马文明时期指代私家财产。在罗马天主教会创造的"patrimonium petri"一词中,集体成为了物质、物品的主人。而法国通过强调国有纪念物的启发意义,认同了集体所有物的观念。公民要通过参观伟大恢弘的纪念物来了解国家的光荣。文化遗产的这种说教式的、雅各宾派(jacobine)的概念,受到强有力的、由专家组成的行政部门的支持,该部门在各个地区执行国家的政治意志。[4]

---

2 详见:Marie Prats:"Les retombées économiques du patrimoine culturel en France", Paris 2011, http://openarchive.icomos.org/1281/1/IV-1-Article1_Prats.pdf (accessed on March 4th, 2016),INSEE:„Tourism in the Ile de France region" http://www.insee.fr/fr/themes/document.asp?reg_id=20&ref_id=23139#inter6 (accessed on March 4th, 2016) Isabelle Backouche:"Mesurer le changement urbain à la périphérie parisienne : Les usages du bassin de La Villette au XIXe siècle". In: Histoire & Mesure, Paris 2010, http://histoiremesure.revues.org/3929 (accessed on March 4th, 2016)

3 Françoise Choay: L'allégorie du patrimoine, Paris 1992, Chapter 3 on French revolution.

4 Philippe Poirrier:"Les politiques du patrimoine en France sous la Ve République. D'une poli-tique étatique à une politique nationale: 1959-2005". In: Maria Luisa CATONI (dir.), Il patrimonio culturale in Francia, Milano 2007, pp. 95-114.

ros.² Seen in retrospect, the way to this success was paved with discrepancy. In the aftermath of the French Revolution, when many castles, churches and artworks fell victim to the destructive fever of the insurgents, the young republic established inventories of endangered heritage. Architectural objects that survived vandalism were classified, measured and restored in the name of public interest, and declared possessions of the French nation.³ The concept of architectural heritage, which in France is called patrimoine, dates back to patrimonium, a term used in Roman civilization to designate assets of a private household. According to the concept of patrimonium petri, coined by Roman Catholic Church, a collective body becomes owner of material belongings. The secular French Republic subscribes to the idea of collective ownership by emphasizing the edifying character of national monuments. Citizens are supposed to learn about the glory of their nation by being exposed to the splendour of memorable objects. This didactic and jacobine conception of patrimoine was supported by a powerful administration of experts executing the political will of the central state in all regions.⁴

The Rotunda of La Villette, a tollhouse designed by Claude Nicolas Ledoux (1736-1806), was commissioned by the ferme générale, a tax perceiving company under France's absolute monarchy. The tollhouse was part of the new city wall, the last and most unpopular undertakings of Ancient Régime, which began only a few years before the Revolution. Ledoux conceived the tollhouses as "Propylaea of Paris", and chose a neoclassical architectural language to endow this design with sobriety and dignity.⁵ The Rotunda of La Villette, which survived revolutionary destruction of the tariff wall, is composed of a central cylinder surrounded by four porticos, and made from undisguised limestone. Ledoux borrowed the geometric principles from a famous precedent: the Palladian Villa Rotonda. Ledoux's design stands as a pivotal point between two major roads leading to Flanders and Germany. Their straight lines define a triangular territory lying beyond the late 18th city limits. Urbanization started here under Napoleon's rule with a new canal infrastructure bypassing river Seine. In front of Ledoux's rotunda, which had already lost its initial function as a tollhouse, the canal Saint Martin is equipped with a large basin where barges could be charged and discharged. The place became

---

2 Marie Prats: "Les retombées économiques du patrimoine culturel en France", Paris 2011, http://openarchive.icomos.org/1281/1/IV-1-Article1_Prats.pdf (accessed on March 4th, 2016),
   INSEE: "Tourism in the Ile de France region", http://www.insee.fr/fr/themes/document.asp?reg_id=20&ref_id=23139#inter6 (accessed on March 4th, 2016)
   Isabelle Backouche: "Mesurer le changement urbain à la périphérie parisienne: Les usages du bassin de La Villette au XIXe siècle". In: Histoire & Mesure, Paris 2010, http://histoiremesure.revues.org/3929 (accessed on March 4th, 2016)
3 Françoise Choay: L'allégorie du patrimoine, Paris 1992, Chapter 3 on French revolution
4 Philippe Poirrier: "Les politiques du patrimoine en France sous la Ve République. D'une poli-tique étatique à une politique nationale: 1959-2005". In: Maria Luisa CATONI (dir.), Il patrimonio culturale in Francia, Milano 2007, pp. 95-114
5 Daniel Rabreau: Claude Nicolas Ledoux, Paris 2005, p. 47 ff.

拉维莱特圆形画廊餐厅（The Rotunda of La Villette）曾经是克劳德·尼古拉斯·勒杜（Claude Nicolas Ledoux, 1736—1806）设计的税关，这一项目由法国君主专制下的一家税收公司"ferme générale"所指派。税关在法国大革命开始前几年动工，是新城墙的一部分，也是封建政权最后且最不受欢迎的一项工程。勒杜将税关看作"巴黎的大门"（Propylaea of Paris），并选择使用一种新古典主义的建筑语言为这个设计注入庄重与威严的意味。[5]拉维莱特的圆形税关在关税壁垒革命所造成的破坏中幸存下来，它由未经加工粉饰的石灰石砌筑，由一根中心圆柱和环绕四周的柱廊构成。勒杜借用了一个著名的古典案例——帕拉第奥圆厅别墅的几何法则。勒杜的设计是巴黎通往佛兰德斯（Flanders）和德国的两条要道上的一个关键点。城市化从这里开始——两条要道在18世纪晚期的城市边界之外划出了一块三角形区域。在拿破仑的统治下，一个新的运河基础设施工程绕过塞纳河展开。在这个早已失去原始收税功能的圆形税关前，拥有巨大河湾的圣马丁运河可供驳船载货和卸货。1867年，巴黎的牲畜市场和屠宰场出现后，这里变成了通往东北面拉维莱特区域的繁忙通道。19世纪后期，在运河和道路基建工程的交汇处，古老城门附近的聚落成长为一个高密度的都市街区。除却功能上的空间复杂性外，这个地方仍然可被辨识为一个拥有特色布局和纪念性圆形税关的城市整体（urban ensemble）。

1902年，这一切发生了改变。高架上的地铁2号线完工，它的铸铁柱正好立在税关南面的柱廊前，勒杜的构图由此被破坏。20世纪60年代到70年代，状况进一步恶化，巴黎东北部发生了经济剧变，许多与牲畜贸易相关的中型企业消失了。现代主义建筑师加斯顿·勒克莱尔（Gaston Leclaire）试图通过一个宏伟的总体规划方案重建这一区域，一个醒目的高层建筑物将替代勒杜的圆形税关。[6]1977年，雅克·希拉克（Jacques Chirac）成为巴黎市长，出于保护建成遗产的目的，这些方案被重新修订。1985年，路易·康的前助手、《今日建筑》（L'Architecture d'aujourd'hui）的主编伯纳德·休特（Bernard Huet）被委派去重新规划这个区域，并重估这个被多种交通流线隔离起来的历史纪念物的价值。休特创造了一片广场空地来强调勒杜的作品，空地的侧面被两个新建的路堤保护起来，使其免受交通喧嚣之扰。这个城市复兴计划恢复了19世纪的城市布局和运河系统的可辨性。[7]在遗产保护政策方面，休特的贡献是具有开创性的。他用一种情境化的手段替代了将纪念物隔离起来的做法。他开创了一种不同历史层和当前需求之间的新对话关系。休特重建了一个有意义和可使用的网络，而不是简单地修复一系列历史建筑，让其无法发挥使用价值。休特的工作于1989年完成，它开创了遗产问题上的新论辩。文化遗产越来越多地被理解为一种传达政治理念的有效途径，以应对本土认同、集体记忆或活态遗产等问题。法国从之前的将遗产看作是一系列杰出的国

---

5 Daniel Rabreau: Claude Nicolas Ledoux, Paris 2005, p. 47 ff.
6 详见：APUR, Starkman, N., Politis, N., (eds). L'aménagement de l'est de Paris. In: Paris Projet: Amenagement, Urbanisme, Avenir. N°28, Paris, 1987: 61.
7 Martin Meade: "Place de Stalingrad". In: The Architectural Review, N° 1111, September 1989, pp. 60-66.

a very busy passage to North Eastern territory of La Villette where the Parisian cattle market and slaughterhouse were implemented in 1867. In late 19th century, the immediate surroundings of the ancient city gate grew into a dense urban neighborhood at the intersection of canal and road infrastructures. Despite the functional spatial complexity, the place was still legible as an urban ensemble with a characteristic layout and the monumental presence of the Rotunda.

Things changed in 1902, when Ledoux composition was impaired by the new Metro Line N° 2 running on top of a viaduct whose cast iron pillars were erected just in front of the tollhouse's Southern portico. And things became worse during the 1960s and 70s when the North-East of Paris underwent a drastic economic change. Many of its medium size enterprises related to the cattle trade disappeared. The modernist architect Gaston Leclaire intended to remodel the territory according to an ambitious master planning scheme, in which Ledoux's rotunda was replaced by a bold high-rise building.[6] In 1977, when Jacques Chirac became Mayor of Paris, the plans were revised in favour of the built heritage. In 1985, Bernard Huet, former assistant of Louis Kahn and chief editor of L'Architecture d'aujourd'hui, was commissioned to redefine the place and to revalue the historical monument, isolated by various flows of traffic. Huet magnified the work of Ledoux by creating an esplanade, which is flanked and protected against traffic nuisances by two newly created embankments. The urban rehabilitation project restored the legibility of the 19th century urban layout and canal system.[7] The work of Bernard Huet was groundbreaking in terms of heritage policy. The architect replaced the isolated consideration of monumental objects by a contextual approach. He initiated a new dialogue between different historical layers and present-day requirements. Huet reinvented a network of significations and uses rather than simply restoring a listed historical monument, which had lost its use value. The work of Huet, which was completed in 1989, inaugurated new debates on heritage issues. Heritage was more and more understood as an efficient means to convey political ideas dealing with local identities, collective memories or living heritage. As a consequence to this diversification, the original French republican understanding of heritage as selection of outstanding national monuments, broke into a multitude of ways to acknowledge material remainings of the past. Particular interests in technical, agricultural or ethnographical themes perpetually diversified the scope of patrimoine, which became definitely a notion of universal presence and signification.[8]

6 APUR, (Atelier Parisien d'Urbanisme), Starkman, N., Politis, N., (eds.): L'aménagement de l'est de Paris. In: Paris Projet - Amenagement, Urbanisme, Avenir. N°28, Paris 1987, p. 61
7 Martin Meade: "Place de Stalingrad". In: The Architectural Review, N° 1111, September 1989, pp. 60-66
8 See also the critique of Marc Fumaroli: "Culture, modernisme et mémoire". In: La Revue des Deux Mondes, Paris, May 1995

家纪念物，转变为认知过去的物质遗存的多种方式。对技术、农业或民族志主题的特殊兴趣，永久地扩大了遗产的范围，遗产由此成为了一个具有普遍存在意义的概念。[8]

休闲娱乐

在距离勒杜的圆形税关不到2公里的地方，矗立着古老的拉维莱特屠宰场。1867—1974年，牲口被聚集在其中央大厅里。这个大厅是方圆55公顷的区域内唯一留存了过往活动痕迹的地方。当弗朗索瓦·密特朗（François Mitterand）于1981年5月当选法国第21任总统的时候，他决定通过建设一个新公园，开创一系列引人注目的"宏大项目"。1982年，一项国际竞赛启动。参赛团队被要求在法国社会后工业状况的背景下，创造一个21世纪的公园。20世纪80年代初，人们拥有大把的闲暇时间，其价值因而下降。休闲活动需要被组织、被建构、被充实。设计应加入休闲娱乐产业，它应创造和满足人们不断更新的需求。竞赛的组织者明确规定，建在旧时劳作地点上的新拉维莱特休闲公园，不应仿造18世纪的如画园景或19世纪干净的城市广场。由于这个未来公园没有历史先例，因此竞赛的简介非常模糊。竞赛留给460个参赛团队很大的发挥空间，他们可以在大片城市空地上大胆地展现对于未来的想法。[9]不幸的是，竞赛发生在一个乌托邦城市项目遭到质疑和嘲讽的年代，经过审议，伯纳德·屈米（Bernard Tschumi）的方案被选中投入建设。

屈米对罗兰·巴特（Roland Barthes）有过仔细的解读，主张在建筑中向个体享乐开放的新享乐主义。在巴特颇具影响力的文章《文本的乐趣》（The Pleasure of the Text）中，他坚持"这就是它（c'est ça）！"的措辞。这种感叹，与其说是解释，不如说是表达一种强烈而不容置疑的肯定。[10]在说出"这就是它！"的瞬间，理性的认可不如令人印象深刻的证明来得快捷。忽然之间，文本和它的读者在欲望的相遇中愉悦地融合在一起。基于巴特的洞见，屈米指向了一个问题：建筑是否能以及如何能激起愉悦的瞬时感受？[11]他通过参加拉维莱特公园竞赛给出了相当简明的答案。屈米拒绝用确凿明晰的结构来填满这片荒地，他也避免使用指示稳定价值系统的建筑象征符号。他的作品更像是建立在一种各向同性的空间观念之上的，其中两条相交的运河几乎不能构成任何特色，边界也只有道路交通系统。公园的表面保持着

---

8   亦可参考：Marc Fumaroli. Culture, modernisme et mémoire[J]. La Revue des Deux Mondes, 1995.
9   Danièle Voldman: "Le parc de la villette entre Thélème et Disneyland". In: Vingtième Siècle, revue d'histoire, n°8, october/december 1985, pp. 19-30.
10 Marielle Macé: "Expérience esthétique et pensée de l'effet, à propos de Barthes". In: Con Roland Barthes, alle sorgenti del senso, P. Calefato, S. Petrilli, A. Ponzio dir., Meltemi 2005.
11 Louis Martin: "Transpositions: On the Intellectual Origins of Tschumi's Architectural Theory". In: Assemblage, No. 11 (Apr., 1990), pp. 22-35.

Leisure

In a distance of less than two kilometres of Ledoux's rotunda lies the ancient La Villette slaughterhouse. The great central hall, in which cattle were gathered from 1867 to 1974, is the last remaining sign of former activities on a vast territory of 55 hectares. When François Mitterand was elected 21st president of the French Republic in May 1981, he decided to inaugurate the impressive series of grand projects with the creation of a new park. An international competition was launched in 1982. Teams were asked to invent a park of the 21st century, taking into account the post-industrial condition of the French society. In the early 1980s, free time has become available in large quantities and its value had fallen accordingly. Leisure had to be organized, structured and filled in. Enter the leisure industry, which knows how to create and satisfy needs that are constantly new. The organizers of the competition made clear that the new La Villette leisure park, which was to be built on a site of labour, should not imitate 18th century picturesque landscape garden nor 19th century hygienic urban squares. Since the park of the future had no historical precedents, the competition brief remained vague. It was left to the 460 competing teams to courageously project their ideas of the future on the vast urban brown field.[9] Unfortunately, the competition was held in a time when utopian urban projections met mistrust and sulphurous critiques. After extensive jury deliberations, the scheme of Bernard Tschumi was chosen for execution.

Tschumi, an attentive reader of Roland Barthes, pleads in favour of a new hedonism in architecture, which is open for individual pleasures. In his seminal essay *The Pleasure of the Text* Barthes insists on the term "c'est ça!", an exclamation that expresses, more than it explains, the intensity of an unquestioned affirmation.[10] In the moment of "c'est ça", rational approbation is shortcut by stunning evidence. Suddenly, the text and its reader are bound to each other in an erotic encounter, a delightful fusion. Based on Barthes' observations, Bernard Tschumi points to the question, if and how architecture can provoke moments of pleasure.[11] The answer he delivers through his competition entry for La Villette is rather laconic. Tschumi refuses to fill the wasteland with clearly determined built structures. He avoids architectural symbols, referring to a stable system of values. His work is rather built upon the idea of an isotropic space, poorly characterized by the crossing of two canals, and fringed by road traffic. The surface of the park remains strangely under-determined. Tschumi's offers an ex-

---

9   Danièle Voldman: "Le parc de la villette entre Thélème et Disneyland". In: Vingtième Siècle, revue d'histoire, n°8, october/december 1985, p. 19-30

10   Marielle Macé: "Expérience esthétique et pensée de l'effet, à propos de Barthes". In: Con Roland Barthes, alle sorgenti del senso, P. Calefato, S, Petrilli, A. Ponzio dir., Meltemi 2005

11   Louis Martin: "Transpositions: On the Intellectual Origins of Tschumi's Architectural Theory". In: Assemblage, No. 11 (Apr., 1990), p. 22-35

奇特的不确定性。屈米的作品提供了一片延伸的平面，就像一张供休闲实验的巨大操作台。他不可思议地拒绝了那种建筑师通常定义开放空间的力量，相反，他选择了一种松散而模糊的布局，让来访者根据自己不可预测的想法来做决定。甚至是他设计的分散在整个草坪上的红色构筑物也基本没有建构起开放空间。它们最多不过是建筑弱点的指示器——无法承载意识形态和功能。屈米刻意将构筑物设计成无用且无意义的记号。它们是不可思议又神秘的物体，只代表它们本身，等待着被好心的游客发现，并加入他们临时的、愉悦的游戏中去。[12]

开园40年之后，拉维莱特公园成为了一个备受称赞的多样化的休闲娱乐区域。公园像马戏团舞台一样容纳了多种多样的项目，没有一个是永久性的。作为密特朗的第一批"宏大项目"之一，拉维莱特公园并没有强调"自然"这一常见的通用的园林景观主题。21世纪的先行者将"项目"的概念提升到了一个普遍性的高度。项目仅允许短期开展。它们为建构未来提供了一个轻捷而可逆的框架。与建成结构不同的是，项目随时间而变。它们可以被重议，可以不断变换。项目是政治工具。它们将个人愿望融入团体活动。今天，拉维莱特管理处的267个职员主要打理公园的项目计划。[13]当地代理商一直在与游客一起，将公园打造成一个可以提供任何可能的娱乐活动的地方。

1987年10月，仅在法国政府签署打造一个比拉维莱特公园大得多的休闲公园的决定性协议7个月之后，拉维莱特公园向公众开放。这个协议的内容是在巴黎东郊方圆20平方公里的区域内建造一个欧洲的迪斯尼乐园。如果我们考虑到休闲娱乐的高政治性甚至普遍性的话，拉维莱特公园和巴黎迪斯尼乐园一定会被认为是有关系的——尽管它们在特点和尺度上都存在着巨大差异。

公众参与

第三个也是最近的一个案例，距离拉维莱特公园3公里，位于欧贝维利耶（Aubervilliers），一个毗邻圣丹尼斯运河和通向佛兰德斯的国道的郊区。这个曾经被称为"Landy"的地方在中世纪的时候曾有市集。19世纪下半叶，工业化进程推进，欧贝维利耶受到外来务工人员和新建工厂的"欢迎"。此外，欧贝维利耶容纳了巴黎市区所排斥的基础设施建设，例如医院、墓地、边防和车库等。

12 Jacques Derrida: "Point de Folie - maintenant l'architecture". In: Bernard Tschumi (ed.) La Case vide, London, 1986.
13 详见：http://lavillette.com/equipes.

tended flatness, a sort of vast operating table for leisure experiments. He strangely rejects the power architects would have to define open spaces. Instead of doing so, Tschumi chooses a loose and vague configuration, which empowers visitors to make their own decisions, to act according to their unforeseeable desires. Even the red follies, which the architect designs and disseminates over the lawn, can hardly structure the open field. They are not more than indicators pointing to the weakness of architecture, which is incapable to bear the burden of ideology and function. Tschumi's follies are deliberately designed as useless and meaningless signs. They are crazy and enigmatic objects standing for themselves, and waiting to be discovered by visitors who might kindly involve them into their instantaneous and cheerful play.[12]

Forty years after its opening, La Villette Park has become a highly appreciated territory for all kinds of leisure activities. Similar to a circus arena, the surface of the park supports a multitude of programs, none of which is supposed to become permanent. As one of Mitterrand's first grand projects, La Villette does not magnify "nature" , the well-tried universal of landscape gardens. The forerunner of the 21st century parks elevates the notion of program to an universal level. Programs allow navigating at short-term. They offer a light and reversible framework for organizing the future. Different from built structures, programs are time-dependent. They can be renegotiated and continually altered. Programs are political instruments. They assemble individual desires within communal events. Today, the 267 employees of La Villette's administration constantly work on the programming of the park.[13] Together with the users, the local agents perpetually transform the park into a territory of possible pleasures.

La Villette Park opened to the public in October 1987, only seven months after the French government signed the decisive contract for the creation of a much bigger leisure park: Euro Disney, which was going to be built on a site of 20 square kilometres at the eastern outskirts of Paris. If we consider the high political, or even universal relevance of leisure, La Villette, Mitterrand's first grand projet and Disneyland Paris must be seen as relatives – despite their obvious differences in size and character.

---

12 Jacques Derrida: "Point de Folie - maintenant l'architecture". In: Bernard Tschumi (ed.) La Case vide, London 1986
13 http://lavillette.com/equipes (accessed on March 4th, 2016)

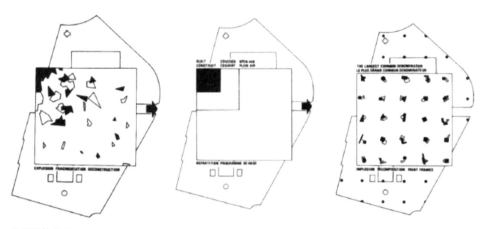

拉维莱特公园
La Villette Park

拉维莱特公园的红色构筑物
Red follies of la Villette Park

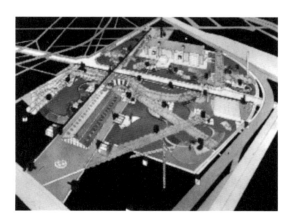

拉维莱特公园
La Villette Park

Participation

The site of the third and most recent example is located three kilometres north of La Villette Park, in Aubervilliers, a suburb adjacent to canal Saint-Denis and the national road to Flanders. In this area, formerly called Landy, fairs were held in medieval times. During the second half of the 19th century, a period of escalating industrialization, Aubervilliers became a welcoming territory for both, foreign workers and new factories. Furthermore, Aubervilliers had to absorb infrastructures, which Paris rejected, such as hospitals, cemeteries, garrisons or garages.

Thomas Hirschhorn is a Swiss artist, born in 1957. For many years, Hirschhorn works in Aubervilliers. He says: "Paris is not Paris without its suburbs. Aubervilliers is a part of Paris. What I need, as an artist, is to live in a space of truth, and this space of truth exists in Paris. As in almost every large city, the space of truth is its suburbs, their so-called banlieues. In Aubervilliers, as in other Parisian suburbs, one can touch the truth and be in contact with it. It's in the suburbs that there is vitality, deception, depression, energy, utopia, autonomy, craziness, creativity, destruction, ideas, young people, hope, fights to be fought, audaciousness, disagreements, problems, and dreams." [14] In 2001, Thomas Hirschhorn was invited by Les Laboratoires d'Aubervilliers, a municipal art agency, to think about a local project. The artist invented the Musée précaire Albinet, an ephemeral gallery, open from March to June 2004, and named after the street where it was built on vacant plot of land. Hirschhorn installed prefab buildings extended by a portico, an atelier for children, a library and a restaurant. The museum buildings were made of wooden slats, which were assembled by layers of carton sealing tape to become the framework for roofs and walls, made of plastic foil and chipboards. Thomas Hirschhorn produces architecture in a way shantytowns are made. All his precarious installations have in common that they are tolerated for an undefined space of time. A precarious building lacks solidity. Its fragile existence can be revoked at any time. The life span of the museum at rue Albinet was fixed in advance at three months, a short duration of intense curatorial work. Every week, an original masterpiece of 20th century art was brought from the archives of Pompidou Centre to the suburbian ephemeral museum. Fifteen people from the neighborhood, between 18 and 25 years old, took part in a training programme. They learned to handle, and to display artworks of famous artists like Kasimir Malevitch, Salvador Dali, Le Corbusier and others.[15] The municipal agency explains that the project allowed sharing the love of art "with people who do not ordinarily have access to

---

14 „Thomas Hirschhorn by Abraham Cruzvillegas". Interview with the artist in: Bomb - Artists in conversation, Issue 113, Fall 2010 (accessed on March 4th, 2016)
15 Denis Moreau: „Le Musée précaire Albinet, Aubervilliers". In: L'Architecture d'Aujourd'hui, N° 354, p. 10-11

托马斯·赫塞豪恩（Thomas Hirschhorn）生于1957年，是一名瑞士艺术家。多年来，他一直在欧贝维利耶工作。他说，如果没有郊区，巴黎就不再是巴黎。欧贝维利耶是巴黎的一部分。作为一个艺术家，他需要的是生活在一个充满事实的空间里，而这个空间就存在于巴黎。正如许多其他特大城市一样，事实空间就是它的郊区，他们所谓的"banlieues"。像巴黎其他郊区一样，一个人在欧贝维利耶可以触碰到事实，并且与之接触。郊区才是拥有活力、欺骗、萧条、能量、乌托邦、自主性、癫狂、创造力、毁灭、想法、年轻人、希望、搏斗、冒险、分歧、问题和梦想的地方。[14] 2001年，赫塞豪恩受到一个名为"Les Laboratoires d'Aubervilliers"的市级艺术机构的邀请，谋划一个当地项目。艺术家创造了一个临时博物馆（Musée précaire Albinet），该馆以其闲置基地所在的街道命名，并在2004年3月到6月间开放。赫塞豪恩装配了一些预制的房屋，并延伸出一个柱廊、一个孩童工作室、一个图书馆和一个餐厅。博物馆由木板条盖成，屋顶和墙壁由塑料片和纸板制成，其主要框架用封装纸箱的胶带层层粘合。赫塞豪恩用搭建临时陋屋的方式建造建筑。他的所有不稳定的装置都因时间的不确定性而被包容。一个不稳定的建筑缺乏固定性，其脆弱的存在随时可能被废弃。这座位于阿比奈特大街（Albinet）的博物馆的"寿命"被事先定为三个月——对于高强度的策展工作来说时间很短。每一周，蓬皮杜艺术中心档案库里就有一件20世纪的艺术杰作被带到郊区的临时博物馆中。来自附近街区的15个18~25岁的年轻人参与了一项培训。他们学习如何处理和展示卡济·马列维奇、萨尔瓦多·达利、勒·柯布西耶等著名艺术家的作品。[15] 艺术机构解释说，这个项目可以让那些"由于社会、经济和文化原因，平时没有机会接触到这些作品的人"分享对艺术的热爱。[16] 倘若没有赫塞豪恩，艺术品收藏家、策展人、赞助商、博物馆参观者和当地居民之间绝不会有这样的接触。当然，这并不能解决阿比奈特大街的严重问题。但是将国家级的艺术瑰宝转移到一个郊区的所谓的危险区域，至少在一定程度上纠正了一些根深蒂固的迂腐成见。赫塞豪恩亲自到博物馆现场，不断作为设计师、建造者、策展人和主持者参与其中的活动，这验证了整个活动真正的参与性本质。

在法国，无论是政治性还是文化性的公众参与都没有真正得到推进。执行团队中的专家和技术官僚构成了这个国家的上层集团，他们仍然坚持雅各宾派的传统，拥护中央集权的决策。在法国，社会想象（social imagination）鲜少能从草根阶层产生。因此，"参与式民主"被社会权力金字塔顶端的国家政治领袖之一塞格林·罗雅尔（Ségolène Royal）推崇，也就不是什么奇事了，她在2007年参加总统竞选时将"参与式民主"作为自己政治路线图的一部分。罗雅尔女士推崇公民论坛和地方委员会的价值，以期加强各个领域内的公众参与度。[17] 尽管罗雅尔最终败选，但许多公民协会由此登上了政治舞台，并且宣称他们作为地方决策的利益

14　详见：Thomas Hirschhorn by Abraham Cruzvillegas. Interview with the artist in: Bomb – Artists in conversation, Issue 113, Fall 2010.
15　Denis Moreau: "Le Musée précaire Albinet, Aubervilliers". In: L'Architecture d'Aujourd'hui, N° 354, pp. 10-11.
16　详见：archives.leslaboratoires.org/content/view/144/lang,en/.
17　详见：http://desirsdaveniruk.canalblog.com/archive/2006/10/23/2977088.html.

these works, for reasons that are mostly social, economic and cultural."[16] The rare encounter of art collectors, curators, sponsors, museum visitors and local residents, which would never have occurred without Hirschhorn, solves certainly not the serious problems of rue Albinet. But the transfer of national art treasures into one of the so-called danger zones of the Banlieue, allowed at least to correct some inveterate clichés. Hirschhorn's presence on the museum's site and his continuous personal involvement as a designer, builder, curator and moderator authentified the truly participatory nature of the whole event.

Participation, be it political or cultural, is not really facilitated in France. The country's elite, employed in executive corps of experts and technocrats, still adheres to jacobine traditions, and advocates centralized decisions. In France, social imagination had rarely grown from grassroots developments. And it was no surprise when even participatory democracy was proclaimed from the top of the social pyramid, by one of the country's leading politician, Ségolène Royal, who put participation on her road map as a candidate for presidency during the 2007 election campaign. Ms. Royal evoked the virtues of citizen forums and local committees to strengthen public involvement in all domains.[17] Although, Segolène Royal lost the elections, many civic associations entered the political arena, and claimed to be heard as stakeholders in local decision-making. Citizens committed to the res publica organized workshops for utopian urban development, pacified the explosive situation in the suburbs, and challenged architectural authorship, which, too often, led to the production of sterile fetishes. Pierre Mahey, one of the French pioneers of participatory architecture, sees the architect as a co-producer. Within a radically participatory design approach, the form-finding processes become a collective undertaking.[18] Mahey arguments remind us Giancarlo De Carlo's saying: Architecture is too important to be left to architects."[19] Participation, as De Carlo understood it, can be seen as a work either for or with users. Working for the users ideally means that architects act as the fiduciary of a common agreement, a binding consensus. Within procedures of the second type, where architects work with users, the consensus remains open and debatable throughout the entire planning process. Participation as a democratic process requires rethinking tools of conception, notations and building procedures in a way that facilitates public discussion.

---

16   archives.leslaboratoires.org/content/view/144/lang,en/ (accessed March 4th, 2016)
17   http://desirsdaveniruk.canalblog.com/archives/2006/10/23/2977088.html (accessed March 4th, 2016)
18   Pierre Mahey: "Le concepteur, cœur de la production du projet". In: L'Architecture d'Aujourd'hui, N° 368, January / February 2007, p. 42-51
19   Giancarlo De Carlo: "Architecture's public". In: Parametro 5, 1970, p. 4-12

相关者,有权发出自己的声音。参与公共事务(res publica)的公民组织了有关乌托邦城市发展的研讨会,平息了郊区一触即发的紧张局势,挑战了建筑设计的作者权威——他们常常创作出了无生气又恋物迷信的作品。法国参与式建筑的先驱之一皮埃尔·马埃(Pierre Mahey)将建筑师看作联合制作人(co-producer)。在一种激进的参与式设计方法的引导下,确定形态的过程成了一项集体任务。[18]马埃的论点让我们想起了吉安卡洛·德卡罗(Giancarlo De Carlo)的话,"建筑太重要了,不能只留给建筑师。"[19]对德卡罗来说,"参与"可以被看作一项为使用者服务或与使用者共同创作的工作。在理想化的情况下,为使用者服务意味着建筑师成为了一个集体认同的信托人。而在建筑师与使用者共同创作的整个过程中,"共识"始终是开放的、可争论的。"参与"作为一个民主的过程,需要概念、标记等反思工具,并需要以促进公众讨论的方式制定程序。

建筑和城市规划作为调解的形式

本文研究的三个案例分别指向目前与遗产、休闲娱乐和公众参与相关的政治考量。这些概念对于20世纪80年代中期至今的社会变化时期极为重要。它们宣告并伴随着对新范式、新价值系统的探索。它们对建筑和政治来说都是一项挑战。在巴黎东北部这个特定背景下的三个案例研究,例证了政治理念对建筑和城市规划的影响。由于与政治直接相关,因此建筑不可能只是呆板地模仿过去,模仿被认可的先例或被奉为典范的模型。在所选案例的情境中,建筑和城市设计根据普遍关心的议题成为改变现有生活世界的一种手段。在三个案例中,建筑被重构成一种调解的手段。伯纳德·休特的设计调解了不同历史层与当前需求之间的关系。他的调解形式脱离了孤立的纪念物本身,使其成为了一个被精心重建的城市整体的一部分。伯纳德·屈米的设计则调解了个人娱乐和集体空间之间的关系。他的调解基于这样一个想法,即把项目视为一种在时间和空间维度上提供可逆结构的方式。托马斯·赫塞豪恩所扮演的调解人,在巴黎城郊安排了一场与现代艺术杰作的相遇。通过他组织的参与性活动,艺术和建筑走下神坛,变得可感、可操作、可理解。建筑和城市规划在意图与期望的网络中展现了它们调解的力量。

(所有图片由作者提供)

---

18 Pierre Mahey: "Le concepteur, coeur de la production du projet". In: L'Architecture d'Aujourd'hui, N° 368, January /February 2007, pp. 42-51.

19 详见:Giancarlo De Carlo. Architecture's public. In: Parametro 5, 1970: 4-12.

Architecture and Urban Planning as Forms of Mediation

The three chosen case studies refer to ongoing political concerns with heritage, leisure and participation. These notions are central to a period of social change, leading from the mid-1980s to the present day. They announce and accompany a search for new paradigms and value systems. They are a challenge for both, architecture and politics. The three case studies on particular situations in the North-East of Paris exemplify the impact of political thought on architecture and urban planning. Architecture in its exposure to politics cannot rely on formalistic imitations of the past, approved precedents or canonized models. Within the chosen situations, architecture and urban design appear as means to alter existing life-worlds according to general concerns. In all three cases, architecture is being reinvented as a means of mediation. Bernard Huet acts as a mediator between different historical layers in their relation to present day requirements. His form of mediation departs form a solitary monumental object, which becomes part of a carefully redesigned urban ensemble. Bernard Tschumi acts as a mediator between individual pleasure and a collective space. His mediation is built upon the idea of the programme as a means to provide reversible structures in time and space. Thomas Hirschhorn acts as a mediator who arranges an encounter with modern masterpieces in a Parisian suburb. Through his participatory event, art and architecture are losing their nimbus and become touchable, manipulable, and graspable. Architecture and urban planning unfold their mediation power within networks of intentions and expectations.

(All pictures provided by the author)

参考文献 Reference:

[1] Marie Prats. Les retombées économiques du patrimoine culturel en France[EB/OL]. http://openarchive.icomos.org/1281/1/IV-1-Article1_Prats.pdf, 2016-03-04.

[2] Isabelle Backouche. Mesurer le changement urbain à la périphérie parisienne: Les usages du bassin de La Villette au XIXe siècle[EB/OL]. http://histoiremesure.revues.org/3929, 2016-03-04.

[3] Françoise Choay. L'allégorie du patrimoine[M]. Paris: Éditions du Seuil, 1992.

[4] Philippe Poirrier. Les politiques du patrimoine en France sous la Ve République: D'une poli-tique étatique à une politique nationale, 1959-2005[M]//Maria Luisa Catoni. Il patrimonio culturale in Francia. Milano: Electa, 2007: 95-114.

[5] Daniel Rabreau. Claude Nicolas Ledoux[M]. Paris: Monum- Éditions du patrimoine, 2005: 47.

[6] Martin Meade. Place de Stalingrad[J]. The Architectural Review, 1989(1111): 60-66.

[7] Danièle Voldman. Le parc de la villette entre Thélème et Disneyland[J]. Vingtième Siècle, revue d'histoire, 1985(8): 19-30.

[8] Marielle Macé. Expérience esthétique et pensée de l'effet, à propos de Barthes[M]//P. Calefato, S. Petrilli, A. Ponzio. Con Roland Barthes, alle sorgenti del senso. Roma: Meltemi, 2005.

[9] Louis Martin. Transpositions: On the Intellectual Origins of Tschumi's Architectural Theory[J]. Assemblage, 1990(11): 22-35.

[10] Jacques Derrida. Point de Folie: maintenant l'architecture[M]/Bernard Tschumi (ed.). La Case vide, London: Architectural Association, 1986.

[11] Denis Moreau. Le Musée précaire Albinet, Aubervilliers[J]. L'Architecture d'Aujourd'hui, 2006(354): 10-11.

[12] Pierre Mahey. Le concepteur, cœur de la production du projet[J]. L' Architecture d' Aujourd' hui, 2007(368): 42-51.

[13] Roland Barthes. Le Plaisir du Texte[M]. Paris: Éditions du Seuil, 1973.

[14] Axel Sowa. Participation in France[C]. K. Geipel ed. Public spheres: who says that public space functions? Hamburg, 2007: 266-269.

[15] Bernard Tschumi. Architectural Manifestoes[M]. London: Architectural Association, 1979.

[16] APUR, (Atelier Parisien d'Urbanisme), Starkman, N., Politis, N. (eds.), L'aménagement de l' est de Paris[J]. Paris Projet - Amenagement, Urbanisme, Avenir. N ° 28, Paris, 1987.

[17] L'Architecture d'Aujourd'hui, N° 225 "Concours" [A], La Villette competition[C]. Paris, 1983.

自然和文化的共生：
土耳其博德鲁姆德米尔别墅群

[土耳其]森古·奥依门·古尔
译者：高长军（英译中）
校译：周梦莹

引言

从现代主义建筑起，不胜枚举的建筑作品已经昭示了新的建筑形式如何飞速遍布全球。这些建筑可以从体量、形式或构造等不同层面来辨识。当然，这既不是说已知形式和格局的组合缺乏创造性，也不意味着前所未有的形式必然更具美学价值。实际上，模仿一直是建筑创作的一个工具。对大多数建筑师而言，已经付诸实践的组合和形式是安全可靠的选择。有些时候，别出心裁的组合可以显得极富创造力。但是显而易见的是，由于当今的全球化大背景，每一个创新很快就会过时。在进行设计时，我们很难不被从他人的奇思妙想和悉心策划得来的视觉经验所影响。因此，真正的追根溯源非常困难。所以，基于对场所精神的回应而萌生的创造力是进行突破的不二选择，否则我们的建筑学设计将会很快陷入雷同。[1]

本文介绍了一个土耳其的案例，此例对场所精神的回应令人信服，同时充分体现了土耳其建筑师图尔盖特·坎瑟沃（Turgut Cansever）深刻的建筑学思考。

坎瑟沃毕生专注于研究，经历丰富，对建筑和艺术等不同领域影响深远。他在艺术领域颇有建树，举办过个人画展，在建筑理论和实践方面也著述颇丰，甚至还擅长演奏芦笛。坎瑟沃出生于一个奥斯曼精英家庭，接触了很多晚期奥斯曼哲学家的著述，包括伊本·阿拉比（Ibn Arabî）、埃尔姆利·哈姆迪·亚兹尔（Elmalılı M. Hamdi Yazır）、欧玛尔·海亚姆（Ömer Hayyam），等等。他在父亲的图书馆中广泛涉猎了赛达德·哈吉·埃尔代姆（Sedad Hakkı Eldem）、恩斯特·迪耶兹（Ernst Diez）、马兹哈尔·赛弗克特·易卜欣尔格伦（Mazhar Sevket Ipsiroglu）等人的著作，这在他成长的过程中起到了重要的作用。[2-4]坎瑟沃与埃尔姆利和亚兹尔的现场对谈对他的发展影响重大；而曾指导他在遗传学和心理美学领域的论文的艺术史学家恩斯特·迪耶兹对他的发展也有重要影响。坎瑟沃还反复阅读了

Nature and Culture Symbiosis: Demir Houses, Bodrum, Turkey

Şengül Öymen Gür

Introduction

It has already been proven from a rich sample of architectural works how new architectural forms float all over the world starting with Modern Architecture. They can be identified at different levels such as mass, formats and tectonics. This is neither to say that combinations of known forms and formats are of lesser level of creativity nor to imply that unprecedented forms are necessarily more aesthetic. In fact, mimesis has always been a tool of creation in architecture. Previously practiced configurations and forms have always been reassuring for most practicing architects. Sometimes extraordinary combinations can be very creative, indeed. However, it is quite obvious that in our era globalization wears out every creative solution and innovation in no time. It is almost impossible to design without being influenced by our visual experiences of someone else's design genius and mastermind. It has become very difficult to trace the original. For this matter, single outlet for creativity emerges from the response to the spirit of the place. Otherwise we are moving fast towards a common architecture [1].

In this article, I will introduce a case from Turkey, which I believe is a convincing example of a response to the spirit of the place, imbued with deep architectural considerations of a unique Turkish architect, Turgut Cansever.

Turgut Cansever-the Turkish architect (1921-2009), owing to his concentrated inquiries and encounters, and his beneficial guidance has contributed to different domains of art and architecture. He did a dissertation in arts, exhibited his drawings, wrote books on the theory and

西方19世纪晚期到20世纪早期研究"存在主义"学说的最主要哲学家的著述,包括"新本体论"的奠基人尼古拉·哈特曼(Nikolai Hartmann)和"过程哲学"的重要奠基人阿尔弗雷德·诺思·怀特海德(Alfred North Whitehead)的著作。他甚至对持有反对论点的尼采(Nietzsche)也有所了解。

"怀特海德寻求一种对于世界的全面的、综合的宇宙观,这种宇宙观提供了一种系统性的、描述性的理论,可被用以解释人类通过伦理、审美、宗教和科学理论而获得的多样的直觉,而不仅仅是科学。"同样地,坎瑟沃基于对《甫苏思》(Fusûs)[2]的理解,融合了奥斯曼的伦理学与美学思想,尤其是伊斯兰哲学,构想出一个独特的复合宇宙观。《甫苏思》是一本论述伊斯兰哲学(本体论)的著述,讨论了"存在的本质",由伊本·阿拉比在1165年所著。这本书认为,宇宙万物都包含一个不变的本质,这就是真理。所以,世间万物应当从其内部进行阐释,通过其与真理(也就是安拉)的内在关联来理解。[3]

在其设计创作过程中,坎瑟沃追求社会的基本理念,以期指引社会团结(social solidarity),最终反映在文化艺术和建筑的本质上。在伊斯兰观念的影响下,坎瑟沃用一种全面统一的观点看待自然、生命、文化、艺术和建筑。根据伊斯兰苏菲派(al-Sufiyyah)的观点[4],现实的中心是安拉,在安拉的面前人是无能为力的。在充分理解新本体论的基础上,坎瑟沃认为,"通过不同层次的方式,万物可以与神性联通;这些方式有诸如材料的、器官的、智力的和精神的,等等。"在这个基本理念下,存在与真理之间建立了一种多层次的联系,每一种联系都可能拥有多重意义。因此,人类是一种无止境的过程状态,一种存在于时间和空间中的客观的动态过程。所以,应该如此理解:建筑是一套准则,设计了涵盖所有万物与真理的关系的秩序。文艺复兴实际上忽略了物质的完整性。[5]

在伊斯兰信仰中,时间和空间是紧密联系、相互作用的两个范畴,涵盖了除安拉以外的一切。因此,运动也是伊斯兰观念的一个核心元素。动态过程即存在,这是至关重要的概念。[6]这与文艺复兴基于——从一个点观看可以得到物体整体——的建筑概念相抵触。

在伊斯兰的空间观念中,最核心的并不是空间的形式,而是建筑元素的各种性质、距离和方向。简而言之,就是人们如何意识到他们自己,并在永恒的空间中徜徉。在伊斯兰的观念中,一个场所作为统一的历史和自然的一部分而存在。建成环境在空间中限定了一个场所,

---

1 参见 https://en.wikipedia.org/wiki/Process_philosophy。"过程哲学"(或者叫作变化本体论)通过变化和发展定义形而上的实体。从柏拉图和亚里士多德的时代开始,哲学家们就普基于不变的物质基础,假设真正的实体是永恒的,而过程从属于永恒的物质。如果苏格拉底变了,生病了,苏格拉底依然是他本身(苏格拉底物质客体没有发生变化),这个变化(病患)仅仅是滑过他的物质客体的一个意外而已:变化是偶然的,反之物质是永恒的。因此传统的本体论不认为实体会发生变化,变化仅仅是偶发性的,而不是必要的。经典本体论让一种知识和其理论变得可能,即变化中事物的学问是不可能掌握的。
2 英文名译作 The Bezels of Wisdom,故又译作《智慧晶棱》——译者注。
3 参见 https://tr.wikipedia.org/wiki/Fusus%27%C3%BCl_Hikem。
4 苏菲派,伊斯兰神秘主义派别的总称,亦称苏菲神秘主义,主张禁欲苦行,潜心修行——译者注。

practice of architecture and even whiffled ney (reed flute). Having descended from an elite Ottoman family he had easy access to books on the late Ottoman philosophy such as İbn Arabî, Elmalılı M. Hamdi Yazır, Ömer Hayyam and so on. His encounters in his father's library with books such as those written by Sedad Hakkı Eldem, Ernst Diez, Mazhar Şevket İpşiroğlu is also worth mentioning in his making of self [2-4]. His live dialogues with Elmalılı and Yazır, as well as the influence of art historian, Ernst Diez, who tutored and supervised him on genetic and psychological aesthetics in his dissertation are important involvements in his development. He had read over and over the leading Western philosophers of "Being" of the late 19th century and early 20th century such as Nikolai Hartmann, the founder of New Ontology, and Alfred North Whitehead, the important founder of Process Philosophy[1] and was even aware of Nietzsche, the dissident.

"Whitehead sought a holistic, comprehensive cosmology that provides a systematic descriptive theory of the world which can be used for the diverse human intuitions gained through ethical, aesthetic, religious, and scientific experiences, and not just the scientific". Likewise, Cansever devised a unique integrated cosmology, compounded of Ottoman ethics and aesthetics, and especially of the Islamic philosophy, *Fusûs*. *Fusûs* (Fusûs'ül Hikem) – "The Essence of Being" is an Islamic philosophy (ontology) book written by Ibn Arabî in 1165, professing that "each being in the universe contains an unchanging essence which is the truth. Hence each being should be interpreted from within, as it is, and comprehended via its interconnectivity with this truth-the God".[2]

In creating his designs, Cansever pursues the essential philosophy of society, which guides social solidarity and eventually reflects in the essence of arts and architecture of that culture. Cansever conceives nature, life, culture, art and architecture in a holistic way shaped by Islam. According to Islamic Sufism, in the centre of reality stands the god and human before the god is incompetent. Having digested the new ontology Cansever argues that "being is connected to god by means of different layers; material, organic, intellectual and spiritual". In this essential sense, being establishes multi-layered relations with the truth, each of which may have multiplicity of meanings. Therefore, human being is in a state of endless process-a virtual

---

1 URL-1: https://en.wikipedia.org/wiki/Process_philosophy: "Process philosophy (or ontology of becoming) identifies metaphysical reality with change and development. Since the time of Plato and Aristotle, philosophers have posited true reality as "timeless", based on permanent substances, while processes are denied or subordinated to timeless substances. If Socrates changes, becoming sick, Socrates is still the same (the substance of Socrates being the same), and change (his sickness) only glides over his substance: change is accidental, whereas the substance is essential. Therefore, classic ontology denies any full reality to change, which is conceived as only accidental and not essential. This classical ontology is what made knowledge and a theory of knowledge possible, as it was thought that a science of something in becoming was an impossible feat to achieve".

2 Fusus'ül Hikem[DB/OL], https://tr.wikipedia.org/wiki/Fusus%27%C3%BCl_Hikem

理解这个空间需要上下观看，体验其内部，而不是文艺复兴式的。任何建筑的研究只可能建立在对运动的追溯上；观察人们进入一个空间并在其中漫步。一个运动的人的经验包括一系列的活动、停止、驻足观赏，以及他对之前有关空间的认知和记叙。坎瑟沃这样描述一个运动的人的状态，"……作为他所面对的所有事件的总体，他在之前记忆和经验的基础上理解建筑元素。"[7]

坎瑟沃将他的建筑作品定义为"一次对映射于形式上的认知的尝试"。他为后世留下了很多可供思考的经典，下面就是其中一个案例。

德米尔假日别墅

科斯拉（Khosla）为促使坎瑟沃的这个项目获得阿卡汗奖（Aga Khan Awards）（坎瑟沃的第三个阿卡汗奖），在1992年对德米尔别墅项目做了一个全面的报告。这份报告是这个项目最可靠的资料来源，以下所提到的数据均基于这份研究。

德米尔假日别墅群包括35个分散的别墅，是一个规模很大的开发区的一期工程（图1，图2）。这些别墅都是为土耳其的中产阶级特别定制的。它们共同代表了该地区迄今为止最有意义的建筑开发，若非如此，这里将会挤满形式上毫无感觉、模式上千篇一律、场地狭小的住宅，然后卖给不幸的客户作为度假别墅。[8]

别墅群位于博德鲁姆的曼德拉亚湾地区（Bodrum-Mandalaya Bay），这里非常接近爱琴海和地中海交汇的地方。博德鲁姆古时候的希腊名字叫作哈利卡纳苏（Halicarnassus），是这个区域最主要的中心。这里的聚落多为希腊风格，也能找到拜占庭风格和奥斯曼风格的痕迹。漫长的夏季从4月一直持续到11月，地中海温暖的海水，丰茂的红松林、橄榄林和柠檬园，使这里成为度假胜地。不过，过去40年旅游的发展带来了无节制的开发，给博德鲁姆造成了很多问题。新建的高速公路改变了这个地方，地价急遽上涨。

别墅群仅占地2.7公顷，坐落于一个面积50公顷、未整体开发的场地内。这块场地位于大海和一片近700公顷的松林之间。别墅群沿着用地的南侧边缘铺开，朝着松林的方向向上逐步抬升。在别墅群的后面，便是陡峭的山坡。

movement- in space/time. Therefore, s/he should be comprehended as such. "Architecture is a discipline, which designs an order to encompass relations between the being and the truth. Renaissance has missed this holistic character of being" [5].

"In Islamic belief, space/time categories are two interlocked categories which bear everything except God. Therefore, movement is also an important factor in Islamic thought. Discerning of the being in the process of movement is essential." [6] This contradicts with the Renaissance conception of building predicated upon the notion that "looked at from a standpoint an object can be conceived in totality".

"In Islamic conception of space, the target is not the form of that space but the various properties of architectural elements, distances and orientations, in short, how human beings perceive themselves and move around in that eternal space. In Islam, a place exists as an integral part of history and nature. Man-made constructions that define a place in the space are conceived by looking upward, downward and experiencing the interiors, unlike Renaissance. Any architectural investigation is only possible by tracing the movement; observing people walking in and around the place." The experience of a moving man consists of a series of activities, stops, viewpoints, as well as his cognitive account of previous involvements with spaces. Cansever describes the life of a walking human, "as the totality of all events he was confronted with. He understands the architectural elements based on the ground of his former remembrance"[7].

Turgut Cansever defined his architectural works as "an effort to reflect cognition onto forms" and produced them for coming generations to appreciate and rejoice. Here is one example.

Demir Holiday Houses

Khosla has documented an extensive report on Demir Houses Project in 1992 to promote Cansever's project for Aga Khan Awards, which Cansever won (his third Aga Khan award). This report is still the most reliable source of information about this project. The following data is based on this study.

Demir vacation houses consist of 35 detached villas which form the first phase of a much

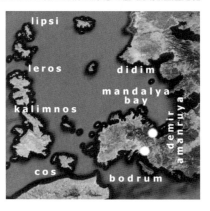

| 1 | 2 |
| 3 | 4 |

1 对文化价值的回应：尊重私密性，博德鲁姆德米尔建筑群
2 对场地的回应和对当地建筑的学习
3 对历史的重新解读：一个小规模的奥斯曼广场
4 德米尔别墅群位于博德鲁姆曼德拉亚湾地区

1 Response to cultural values: guarding for privacy, Demir Houses, Bodrum
2 Respose to the site and references to local architecture
3 Re-interpretation of history: a small size Ottoman piazza
4 Demir Houses on the jutting Bodrum peninsula in Mandalaya Bay

方案

以下资料同样从上述科斯拉的报告中选取。在1971年最初的设计版本中，别墅有3~5个床位的户型，并在客厅内设有1个额外的床位。当时提供的用地面积取决于所选别墅的面积，别墅面积从120平方米到300平方米不等。在坎瑟沃的记述中，单层和双层别墅混在一起，分布在不同的水平面上，以实现"多样、活泼的外观和建筑群的统一性"。

该设计概念的价值在于不仅回应了用户的需求，同时在建造上取得了很好的效果。这些都与坎瑟沃独特而全面的建筑观紧密相连，是其逐渐发展并持续坚持的态度。在设计的演变过程中，以下几个不可或缺的方面也塑造了方案的特性：

larger development (fig.1, fig.2). The houses have been custom designed for middle-class Turkish families. "Collectively they represent by far the most significant architectural development in the region which is otherwise inundated with insensitive, repetitive house types crowded onto small sites and sold to hapless buyers as vacation houses." [8]

The houses are located in the Bodrum-Mandalaya Bay Region, very close to Bodrum where the Aegean Sea meets the Mediterranean Sea. Bodrum is the main focus of the region whose ancient Greek name was Halicarnassus. Settlements in this region have Hellenistic, Byzantine and Ottoman traces. The long summer season from April to November, the warm waters of the Mediterranean Sea and the presence of red pine forests, olive groves and citrus orchards, make the region exceptional for tourism. The development of tourism in the region over the last forty years has begun to pose severe problems due to uncontrolled development, though. New highways have transformed the region and the land values are spiralling upward.

The project under review measures only 2.7 ha and sits within an overall undeveloped site of 50 ha, which is set within a pine forest of 700 ha and the sea. The 2.7 ha that comprises the built portion is located along the southern edge of the larger undeveloped site and slopes gently upwards towards the pine forest. The land then rises steeply toward the hill.

The Programme

Below information is also excerpted from Khosla's above mentioned report: The villas that were offered in the original offer of 1971 had 3, 4 or 5-bed units with an extra bed in the living room. The plot sizes offered at that time were dependent upon the size of villa selected and range from 120 m^2 to 300 m^2. Single and double storey villas were to be mingled and located at different levels in order to achieve "a varied, lively appearance as well as architectural unity", as Cansever notes.

The evaluation of the design concept, its response to user requirements and the excellent standard of construction are all linked to the unique and comprehensive view of architecture that the architect Turgut Cansever has continuously developed. Integral to the evolution of the design, the following aspects of the project also define its unique character (Ibid.):

（1）丰富的形式和体量。

（2）这35栋别墅每栋都有可观海景的视角。每个别墅的主人都可以在建造之初选择自己希望建造的位置。当然，自由选址的唯一限制，就是不能阻挡到其他已经建好的别墅观海的视线。

（3）材料的统一性。整个方案采用了一套统一的建筑材料语汇，选材包括石材、木材和少量的清水混凝土。

（4）对环境的考量（对树木、土壤和海水的保护）。不能砍伐任何一棵树；汽车被尽可能地排除在场地之外；保护整个海岸线，避免污染海水。

阐释

与自然的对话

自然是建筑学理论与实践中最复杂的概念之一。它在建筑学中有着丰富的演绎，也在很大程度上充满了争议。同时，自然又是一个重要的、有着精神内涵的物质实体。在柏拉图之后西方的建筑学观点中，通过富有创造性的模仿生成形态一直是形态学的一条原则。自古以来，大自然中的秩序、比例、和谐和多样性促使了建筑设计原则和符号的产生。这些原则和符号在西欧建筑学风头正劲的时候被固化，又在现代主义建筑兴起时实现了国际化。

近些年来，自然环境在城市建设和发展的过程中受到了压制。因此，"生态建筑""地域性"和"触觉"等概念在建筑评论领域逐渐浮现。但在伊斯兰概念中，自然是不能被贬低或者无视的重要实体。德米尔别墅群项目提供了极好的自然体验和场所感。坎瑟沃在自然面前表现出来的谦逊和克制及其朴实无华的性格，帮助他达到了这一效果。他并没有改造甚或重塑自然，而是保留原始的自然状态。对他而言，自然并不是对建筑的完善和补充，而是建筑整体的一部分（图3—图6）。

与环境的对话——"有机建筑"

"有机性"是广为流传的建筑术语之一，常常跟现代主义建筑大师赖特（F. L. Wright）[9]联系在一起。赖特对这个概念做出了最好的诠释，他认为建筑是基于其内在的属性[5]，由内而外

5　作者将其称作"DNA"——译者注。

- Variety of form and massing.

- Each of the thirty-five villas has a view of the sea. Each of the owners could select the spot at which he wanted his villa to be built. The only constraint in his freedom of choice of location was that his type of house should not obscure an existing villa's view of the sea.

- Unity of materials. A common architectural language was evolved for the entire construction consisting of stone and wood and limited use of exposed concrete.

- Environmental concerns (conservation of trees, soil and sea). Trees could not be cut! Cars were excluded from the site as much as possible. The shoreline was protected and the pollution of the sea was avoided.

Interpretation

Discourse with Nature

Nature is one of the most complex concepts that exist in the literature on architecture; and its interpretation in architecture is rich and largely controversial. It is at the same time both a physical reality and vital essence. In Western architecture, after Plato, it has been a creative morphological principle that shapes form via mimesis. Order, ratios, harmony and variety found in nature have led to the formation of architectural design principles and symbolism since archaic times, have been solidified during powerful periods of Western European architecture, and have been internationalized during Modern architecture.

Recently, it is argued that nature has been suppressed in the process of city building and development. Therefore, concepts such as "ecological architecture", "the role of place" and "the tactile" surfaced in the criticism of architecture. However, nature is not an entity to denigrate or downplay in Islamic conception of space. In Demir Houses, Cansever provides an astonishing experience of nature and sense of place. His modesty and shyness before nature and his naturalness are his instruments to reach this goal. He does not trim or re-shape the nature. He leaves it as it is because to him nature is not something that completes architecture but indeed an integral part of it (fig.3-fig.6).

5 德米尔别墅群局部平面
6 大自然前的克制
7—10 博德鲁姆地区建筑案例
5 The partial layout of Demir Houses
6 Shyness before nature
7—10 Examples of local architecture in Bodrum region

5 6

7—10

Discourse with the Setting / "Organic Architecture"

The term "organic" which is one of the broadest concepts in architecture is frequently associated with Modern architect F. Ll. Wright who had rendered the best explanation for the concept by arguing that a building grows from inside out based on its interior programme – the plan[3] and like any other piece of natural organism it spreads over the pleats of the topography[9]. He had advocated his treatise as an opposition to the academic classicism of the Beaux Arts. But the poor and superficial views of the late 20th century regarded it as curved and flamboyant forms simulating the nature-a kind of mimesis. In Wright's original understanding it is a reverential response to the site, function, materials and other contingent qualities of the surrounding, in such a way that the result at once becomes organic and place-specific. In this sense Cansever's project is at the same time both organic and imbued with profound sense of place.

Sense of Place (context & history)

Cansever harmonizes the architectural style with the natural and cultural background of the surrounding region. The domestic architecture of the region consists of simple repetitive modular blocks in stone, punctured with small rectangular windows. The Aegean vernacular houses typically have flat roofs constructed in timber and floored at two or three levels. Cansever employs motives which resemble endogenous style of the geography. With interpreted façades and tectonics such as chimneys and oriels Demir Houses refer to the Bodrum houses (fig.7-fig.10).

He also stylizes old Ottoman and Seljuk urban motives. The general layout of the setting is designed to provide places of contact and congregation which is very important in Ottoman urban design. Ottoman neighborhood centers do not display the fact that there is a mastermind behind them. They look as if they simply happen to be there, as if it were their fate. They are very organic; open to changes and additions. Likewise social meeting places competently designed by Cansever exist as if they just happen to be there by chance, or by coincidence. In so doing he successfully maintains equilibrium of movement and tranquility; and triggers memory.

Response to Function and Cultural Values—Privacy

In this sense Cansever's design in addition to being a terrific response to site, nature and history is also competent in caring for function and human relations. Although plans tend

---

3 I call this DNA in my lectures.

生长出来的，像其他任何自然生物一样遍布在起伏的大地上。赖特的理论是对布扎体系学院派建筑的一种反抗。20世纪晚期的一些观点对赖特理论的理解过于肤浅，仅认为其是通过弧线和夸张的自由形态对自然的一种模仿。赖特最初的观点其实是对场地、功能、材料和周边其他环境要素的一种尊重和回应。所以，通过这种方式诞生的建筑显得更加有机、有场所感。从这个意义上讲，坎瑟沃的方案同样可称得上"有机建筑"，并具有深刻的场所意义。

场所感

坎瑟沃的建筑风格与自然及当地的文化背景达到了高度协调。该地区的住宅建筑由石材建造，以简单重复的模块构成，在此基础上点缀有矩形小窗。爱琴海当地民居通常是二到三层，用木材建造平屋顶。坎瑟沃的设计初衷类似于基于地域性的内生方式，通过对表皮的塑造和传统建造的再现，如烟囱和凸窗，德米尔别墅群与博德鲁姆地区的住宅取得了风格上的统一（图7—图10）。

坎瑟沃效仿了老奥斯曼风格和塞尔柱城市风格（Seljuk urban style）。在德米尔别墅群的整体布局中，他设计了在奥斯曼城市设计中非常重要的、可以提供交流和集会的场所。奥斯曼邻里空间的形成看起来自然且恰当，仿佛没有设计师的参与，充满了偶然性。这些邻里空间非常有机，亦允许改造和加建。同样地，由坎瑟沃所设计的社交空间也恰到好处，看起来如同自发形成。通过这些策略，坎瑟沃保持了空间的动静平衡，触发了人们的记忆。

对功能和文化价值观——"私密性"的回应

从这个意义上讲，坎瑟沃的设计不仅仅是对场地、自然和历史的一个极好的回应，同时也很好地注意到功能需求与人之间的关系。尽管从平面上看起来这些别墅丰富多彩、差异明显，但是通过对整体布局的控制，它们最终达到了统一性和整体性。传统观念中对邻里的尊重得到了保护；住户们在拥有极好海景的同时，也避免了因开放性和开窗位置而暴露隐私。保护私密性对于这个群体而言是很重要的一个行为原则，所以任何窥到其他住户室内、前后花园，特别是入口的可能性，都通过别墅之间的巧妙组合被规避。

选材和建造

从18世纪中期起，工程技术的发展给建筑师提供了技术上的支持。19世纪末期，美国高层建筑的发展，又进一步推动了现有技术资源的发展。整个20世纪，玻璃的应用呈指数式增长。建筑界出现了"高技派"，金属和玻璃组成的新表皮风格极其简单却倍受青睐。在20世纪末期，信息技术的发展让电脑制图成为了可能，将建筑师从正交结构中解放出来，允许其使用奢侈的自由形态。坎瑟沃至少可以选择现在看起来有点传统、但当时还算前沿的钢筋混凝土，或者干脆使用钢结构。但由于该地盛产石材，过去的建设几乎都使用石材完成，安纳

towards diversity and differentiation, unity and integration of the setting is maintained by the manipulation of the layout. Traditional respect for the neighbor is guarded: neighbors while having a magnificent view of the sea are protected against other's gaze owing to the location of openings and windows. Privacy is an important cultural paradigm for this population. The possibility of having a glance of other user's interiors, back and front gardens and especially the entrances are circumvented by locating houses in a clever way to avoid this issue.

Choice of Materials and Techniques

Engineering provided architects with technical support since the mid of the 18th century. Towards the end of the 19th century, high buildings stretched the resources of existing technologies in America. Use of glass grew exponentially during the 20th century and its refinement in architecture came to be known as "high-tech". Metal and glass surfaces the new style favoured were exceedingly simple. At the end of the 20th century, information technology made possible computer drafting and liberated designers from the orthogonal nature of construction and have allowed the use of arbitrary and extravagant forms. Cansever could at least have chosen –now conventional- the use of reinforced concrete or steel. However, in this region the local construction in the past was built almost invariably in quarried stone, which is abundant throughout the region. Anatolian builders had developed rich patterns of masonry, different vaulting forms using small stones that had been shaped through elaborate geometrical cutting techniques (fig.13-fig.15). Cansever employed different types of local stone to maintain the feeling for locality and elicit precious memories of the past (fig.11).

Sense of Touch

Ocular-centrism is a utilitarian-aesthetic perspective, which has dominated the perception and apprehension of spatial quality and the success of the architectural work in the West over the years[10-11]. Cansever opposed this limited understanding of aesthetics all his life.

We use our senses and instincts in comprehending a phenomenon. In this process, visual stimuli are not more dominant than the auditory, olfactory, haptic and kinaesthetic stimuli [12]. Paterson repeats an old saying: "To see is to believe; but to sense is real…We are sensory homunculus."[13]

All sorts of stimuli require a certain kind of skin; even the cornea of the eye is a very erudite kind of skin. Thus, the sense of touch comprehends entirety of our experiences with the environment[14]. The haptic elements of the object send certain messages to our brain, which

托利亚（Anatolian）的工匠发明了丰富多样的砌石方式。工匠用小石块构筑出不同的拱形结构，这些石块通过精巧的几何切割工艺加工而成（图13—图15）。坎瑟沃采用了多种多样的当地石材以保证方案的地域性，同时激发了人们对过去的回忆（图11）。

触感

视觉中心主义作为一种实用主义审美观，主导了人的知觉和对空间品质的理解。多年来在西方世界，视觉中心主义还影响了建筑作品的成败。[10-11]坎瑟沃毕生都反对这种对美学的局限性理解。

我们在理解一个现象的时候运用我们的感官和直觉。在这个过程中，视觉刺激其实并不比听觉、嗅觉、触觉和动觉更占优势。[12]帕特森引古谚所言："人们习惯于相信眼睛所看到的，但是感知到的才是真实的。其实我们是感知上的侏儒。"[13]

各种刺激都需要特定的介质（skin），即便是眼角膜这样的器官都可以成为一种非常强大的介质。因此，触感实际上包含了我们对环境的全部感知。[14]通过对物体的触觉感知，我们的大脑接受了特定的信息，让我们追忆起过去的记忆、感受和情绪，这些又决定了我们面对建筑和环境时所反应出来的欢喜（kind）和紧张（rigor）（图12）。

提及记忆，本雅明（Walter Benjamin）强调说，"所谓过去，只是被抓住的一个瞬间形象。"[15]让·鲍德里亚（Jean Baudrillard）从另一个方面表示，"图像和记忆，屏幕和真实之间的距离变得难以察觉。所以根本没有办法取得最初的原真记忆，它已经被重塑或者变形。"[16]帕拉斯马（Pallasmaa）最近写道，"我们从视觉上获得的对世界的认知不仅仅是一张图片，而是一系列连续的、可塑的架构，通过记忆持续地整合单个的感知。"[17]

建筑空间的情感作用取决于留在人们脑海中的暗示的深度，有情感的建筑是有触感的。坎瑟沃通过对触觉元素的运用，塑造了一个就体验者而言充满强烈场所感和愉悦感的空间。坎瑟沃使我们想起过去的美好回忆和感知（图16—图23），而这些明显是他已有的。

按照伊斯兰的习惯，坎瑟沃崇尚简单朴素。在环境的各个角落和未来发展的可能中，人体尺度都得到了充分的考量。西奥多·阿多诺（Theodor Adorno）在1962年写道，每一个当代的文化现象，即使一个完整性的模型，都很容易在培养低俗的环境中窒息。[18]跨越了几千年，在这片土地上，德米尔别墅与安纳托利亚传统的建筑完成了一场优雅的、一丝不苟的对话。

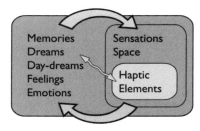

11 12
13-15

11　对历史的追忆：石材
12　触觉体验在我们与建筑和场所之间的地位
13—15　自11世纪起，安纳托利亚的石艺
11　Reminisces of history: stone
12　The role of tactile experiences in our associations with buildings and places
13—15　Stonework dating from the 11th century onwards, in Anatolian

16 用石材建造的德米尔别墅群
17—20 安纳托利亚地区本土传统建筑的凸窗
21—23 土耳其的壁炉
16 Demir Houses built in stone
17—20 Oriels from the Anatolian vernacular and urban traditional
21—23 Some fireplaces from Turkey

16

21—23

17—20

bring back old memories, feelings and emotions that in turn dictate the kind and rigor of our reactions to the buildings and environments (Fig.12).

Concerning memory Benjamin had underlined that: "The past is seized only as an image on a fleeting moment…" [15] Jean Baudrillard on the other hand, had argued that: "The distinction between the image and the memory, between the screen and the real becomes imperceptible. There is no original memory to be retrieved; it has already been re-written and transformed." [16] Pallasmaa recently noted: "Our visually acquired image of the world is not a picture at all, but a continuous plastic contruct that keeps integrating singular perceptions through memory."[17]

The affectivity of architectural space depends on the profundity and deepness of the suggestions left on mind; affective architecture is haptic. Cansever reminds us of our sweet memories and feelings for the past, which obviously he already has, and renders a strong sense of place and pleasure on the part of the observer via haptic elements (fig.16-fig.23).

Cansever is satisfied with the simple and austere, as Islam suggests. Human scale is guarded at every corner of the setting and the vistas offered by the setting. In 1962, Theodor Adorno wrote: "Every contemporary phenomenon of culture, even if a model of integrity, is liable to be suffocated in the cultivation of kitsch."[18] However, Demir Houses are in most refined and meticulous communication with the past architecture of Anatolia, spanning over thousands of years and with the place in which they exist.

参考文献 Reference:

[1] Gür, Ş.Ö. and Erbay, M., Globalization in Architectural Forms: Trifurcation, Shigeru Kuwahara, Japan[M], Liaoning Science and Technology Publishing House, 2011.

[2] Düzenli, H. İ. İdrak ve İnşa: Turgut CanseverMimarlığın İki Düzlemi[M], Istanbul: Klasik Yayınlan, 2009a.

[3] Düzenli, H. İ. Vefeyat: Turgut Cansever[M], İslam Araştırmalan Dergisi, No. 22, 2009b,. 160-181.

[4] Düzenli, H. İ.The Analyses of Form Function Technology and Meaning in Turgut Cansever's Projects in the context of Architectural Autonomy and Civilizational Self-perception[D], Master's Thesis, supervised by Şengül Öymen Gür, KTÜ, Trabzon, 2005.

[5] Cansever, T., Şehir ve Mimari Üzerine Düşünceler, Reflections on the City and Architecture[M], İstanbul: Ağaç Yayıncılık, 1992.

[6] Cansever, T.İslam'da Şehir ve Mimari, The City and Architecture in Islam[M], İstanbul: İz Yayıncılık, 1997.

[7] Cansever, T., Kubbeyi Yere Koymamak, Not Putting the Dome Down[M], İstanbul: İz Yayıncılık, 1997.

[8] Khosla, R. Demir Holiday Villas, Aga Khan 1992 Technical Review Summary[M], 1992.

[9] Wright, F. Ll. Frank Lloyd Wright On Architecture: SelectedWritings 1894-1940[M]. Duell, Sloan and Pearce, 1941.

[10] Erkartal, P. Ö. and Ökem, H. S. The Phenomenon of Touch in Architectural Design and A Field Study for Haptic Mapping[M]. Megaron. 2015, 10(1), pp. 92-111.

[11] Gür, Ş.Ö. Nilgün Kuloğlu et al, Eds. Reading The Place[M],Inaugurating Lecture, Pedestrian Traces in Urban Centres, Proceeding of the Congress, Trabzon Chamber of Planning Seminar& Workshop, Trabzon, March 11-15, 2015.

[12] Gür, Ş. Ö, Deciphering of the Palimpsest: Proposed as a Soft Method for the Information Gathering Phase of the Architectural Design Process[A], 14th Building &Life International Congress (11-12 October), Bursa, Turkey, 2002 (keynote speech).

[13] Paterson, M. The Senses of Touch: Haptics, Affects and Technologies[M]. New York: Berg Publishers, 2007.

[14] Montagu, A., Touching: The Human Significance of the Skin[M], New York: Harper& Row, 1978.

[15] Benjamin, W. trans. by John Osborne, The Origin of German Tragic Drama[M], London, 1985.

[16] Baudrillard, J. trans. by Işık Ergüden, La Transparence du Mal, 1990[A], Kötülüğün Şeffaflığı-Aşırı Fenomenler Üzerine Bir Deneme[M], Istanbul: Ayrıntı Yayınlan, 1995.

[17] Pallasmaa, J. In Praise of Vagueness[J], Space & Psyche, Center 17, Elizabeth Denze & Stephen Sonnenberg, Eds. 2013, 254-270.

[18] Frascina, F. General Introduction[A], Modern Art Culture: A Reader[M], Francis Frascina, Ed. Routledge Taylor& Francis Group, London and New York, 2009, pp. 1-11.

[19] Cansever, T. İstanbul'u Anlamak, Understanding Istanbul[M], Istanbul: İz Yayıncılık, 1998, 2008.

[20] Cansever, T. Ev ve Şehir Üzerine Düşünceler, Thoughts on the House and the City[M], İstanbul: İz Yayıncılık, 1994.

约翰内斯堡的普遍性与特殊性：
迈向再生性发展

[南非] 凯伦·艾克尔
译者：杨柳（英译中）

引言

目前，全世界越来越关注一种系统的可持续性，它超越了绿色建筑，目的在于在再生、互惠的设计基础上，创造健康的城乡生态系统。这种方式不仅包括高端的商业及公共建筑中使用的最新型绿色建筑材料和技术的发展，也包括较小程度的、深思熟虑的干预措施，使得更多小型项目也能产生重要的社会和环境影响。

在南非，有一小群从业者和专业学者走在了从可持续发展转向再生性发展的潮流前沿，他们认为应肩负起个人责任，设计能够催化更宽广的、使周边环境再生的干预措施，以建立能加强社会凝聚力以及能使周边社区健康发展的系统。

修布罗：背景介绍

约翰内斯堡是构成豪登省和都会区的两大城市之一，该省人口超过1.32亿[1]。豪登省是南非目前地域最小的省，却是撒哈拉沙漠以南非洲地区的经济枢纽，吸引着来自全国及非洲其他地区的人群前来寻找工作机会。

本地工作室（Local Studio）为拓展基金会（Outreach Foundation）设计的新楼位于约翰内斯堡历史中心商业区东北部的修布罗。修布罗在19世纪80年代末是一个发展迅速的矿业城

---

1 南非统计局. 统计公报 P0302：2015年中人口估算. 2015年，第2页。

The Universal and the Particular in Johannesburg: Moving towards Regenerative Development

Karen Eicker

Introduction

There is currently a growing focus worldwide on systemic sustainability that moves beyond green building and aims to create healthy urban and rural ecosystems based on a regenerative and reciprocal approach to design. This approach accommodates not only the latest developments in green building materials and technologies utilized in high-end commercial and public buildings, but also the smaller scale, considered interventions that allow more modest projects to have a meaningful social and environmental impact.

At the forefront of this drive from sustainable development towards regenerative development in South Africa is a small group of practitioners and academics who believe in taking personal responsibility for designing interventions that can become catalysts for broader regeneration within the immediate environment, thus creating systems that enable the improved social cohesion and health of the surrounding community.

Hillbrow: A Context

Johannesburg is one of two large cities that constitute the province and metropolitan region of Gauteng, which has a population of over 13.2 million people[1]. Gauteng is by far the smallest

---

[1] Statistics South Africa. Statistical release P0302: Mid-year population estimates 2015, p. 2.

镇，是该市最早确立的郊区之一，此后一直是该市接收外来者的区域，也是该国历史上人口最为密集的高楼住宅区。

20世纪四五十年代，该地区的早期住房被高层公寓楼房取代，以安置第二次世界大战的移民；之后在七八十年代，修布罗发展为市中心最具世界性的区域。根据Planact/CUBES一篇题为《城市用地：约翰内斯堡市的穷人空间？》[2]的研究，尽管在种族隔离制度下，族群住区法明令禁止迁移，还是有很多人从住房资源严重短缺的乡镇迁往工作机会较好的地方。

"这种情况引起房租上涨，因为房东可以利用黑人租户脆弱的黑民身份来盘剥他们，这进而导致转租和过度拥挤的现象，于是楼房情况开始迅速恶化。"

"由于本地区内及附近的交通条件、工作机会较好，并有大量适合不同支付能力水平的住所，修布罗仍然不断地迎来一波又一波的新住户，其中很多来自临近的非洲国家。"

今天的修布罗仍是那些前来这个城市寻找经济机会的人们的落脚点。由于这些人之中有很多是非法入境的，他们没有能让自己获得固定工作或是找到长期住所所必需的身份文件。

住房与城市发展研究所2010年一篇题为《非洲城市化：再访市中心》的报告称，"对于该市未来的很多居民来说，在市中心做'贫民窟房东'或在周边做'棚屋农民'是仅有的较为实际的住房选择……这种不合乎常规的、赖以维持生活的条件，被控制、转租或租用的住所，对于城市的贫困人群来说，是可供使用的最多、最好的机会。"[3]

因此，非正规系统找到了容纳尚未进入正式系统的人群的方式，常常是非法地在一些已被分裂、分割、改变用途的建筑物内容纳这些人群，并且所容纳的人数远远高于建筑物本身及其周边设施原本适合的目标人数。

修布罗的情况就是如此，由于缺乏投资吸引力，房产价值下跌，房产状况恶化，并且很多已被离开的业主/房东遗弃，或被不满或"非法"的房客控制，他们自称为非正式法人团体，向居住者收取租金，却未对这些建筑进行任何管理或维护。结果就是，过去二十年里，该地区犯罪率攀升，尘垢堆积，民用设施失修，社区社会服务水平低下，公共开放空间短缺。

约翰内斯堡住房公司（JHC）认识到这一情况，于2004年在修布罗发起了eKhaya社区改进计

---

[2] Planact. Urban land: space for the poor in the City of Johannesburg?: Summary of findings of a 2007 joint Planact/CUBES study on Land Management and Democratic Governance in the City of Johannesburg[R/OL]. http://www.planact.org.za/wp-content/uploads/2014/08/2.-Urban-land-Space-for-the-Poor-Research-Summary.pdf.

[3] Peter Ahmad. Inner city nodes and public transportation networks: Location, linkages and dependencies of the urban poor within Johannesburg. IHS Working Papers, 2010(26): 11,12.

province in South Africa, yet it is the economic hub of sub-Saharan Africa, attracting many people from around the country and the rest of Africa who are in search of work opportunities.

Local Studio's new building for the Outreach Foundation is located in Hillbrow, northeast of Johannesburg's historic central business district. Established as one of the first suburbs of the city when it was just a burgeoning mining town in the late 1880s, Hillbrow has always been a reception area for newcomers to the city and is historically the most densely populated high-rise residential area in the country.

In the 1940s and 50s, the area's early houses were replaced with high-rise blocks of flats to accommodate World War II immigrants; while later, in the 1970s and 80s, Hillbrow developed into the most cosmopolitan area of the inner city. According to the Planact/CUBES study entitled "Urban land: space for the poor in the City of Johannesburg?"[2], despite the prohibitions of the Group Areas Act under Apartheid, many people began to move away from the townships where there were severe housing shortages to areas that were better located in terms of work opportunities.

"This situation caused rents to increase because landlords could exploit the vulnerable status of illegal black tenants, which led to subletting and overcrowding, and as a result, buildings began to deteriorate rapidly.

"Because of good transport links, job opportunities in and around the area, as well as an array of accommodation at various affordability levels, Hillbrow continues to experience waves of new residents, many from neighboring African countries."

Today, Hillbrow remains a landing place for people migrating to the city in search of economic opportunities. As many of these people have entered the country illegally, they do not have the necessary identity documents that would enable them to seek permanent employment or find permanent accommodation.

The Institute for Housing and Urban Development Studies' 2010 report entitled "Urbanising Africa: the city centre revisited" states, "For many of the city's prospective residents, inner-city opportunities facilitated by 'slum-lording' or peripheral opportunities via 'shack-farming' are the only tangible housing options… the greatest number and choice of opportunities

[2] Planact. Urban land: space for the poor in the City of Johannesburg?: Summary of findings of a 2007 joint Planact/CUBES study on Land Management and Democratic Governance in the City of Johannesburg[R/ OL]. http://www.planact.org.za/wp-content/uploads/2014/08/2.-Urban-land-Space-for-the-Poor-Research-Summary.pdf.

划——在高人口密度、以居住为主的市中心社区进行的首个此类计划。过去十年，eKhaya与业主、建筑管理者、城市机构、居民和其他临近的社区组织建立了合作关系。

如今，该计划相当成功地改善了该区域物理环境的总体质量，尤其是在清洁、安全、维护、公共空间的升级改造方面；并吸引了私有投资和公共投资前来此地。尽管所采取的干预措施往往是针对城市管理和设施管理的，却极大地改善了在该地区居住和工作的人们的生活。

拓展基金会

拓展基金会大楼位于eKhaya社区商业区，新楼供eKhaya及社区内其他活跃机构用于召开会议。该项目除了在2015年赢得豪登省建筑协会奖（GIFA），同年还获得圣戈班南非社会贡献奖（建造类），此前还在2014年南非钢铁奖活动中受到嘉奖。

路德团体拓展基金会（The Lutheran Community Outreach Foundation）作为一个组织，通过技能发展利用富含艺术的项目鼓励创意旅途，为居住在约翰内斯堡市中心的人群提供支持。据该基金会称，"城市退化和失业不断引起民众对周围环境的愤怒和不满；而且……年轻人……希望向同龄人和成年人寻求引导，但这些人实质上却不甚关心。"

"各种各样的项目给孩子、年轻人和成年人提供了参与到与艺术、文化和传统遗产相关的活动中去的机会，这些活动有助于促进沟通、参与以及社区建设，因而也能促进居民与年轻人之间的互动，并在学校假日期间，为孩子们提供新的活动选择。通过团队建设、自我价值建设和身份探索，拓展基金会给众多家庭提供了一个全面的方式，促进了这些社区之间的沟通。"[4]

这个项目启动两年之前，该基金会获得了国家彩票拨款，用以进一步推进其目标。他们希望建立一个场所，把他们的数个拓展项目集中起来，包括一个定期进行的舞蹈计划、音乐计划、计算机学习中心、马林巴琴乐队，以及一个咨询处，并开放项目地点已有社区中心的内部空间作行政用途。

最初项目地点选在19世纪90年代末的原路德教堂，那里至今仍然留存着老教堂。20世纪60年代，德国领事馆捐款扩大这个场所，希望能纳入一个社区中心，但这个建筑项目一直未能完工。

---

4 拓展基金会 http://outreachfoundation.co.za/index.php/about/vision-mission.

are available to the urban poor in irregular, subsistence-based arrangements in hijacked, sublet or rented accommodation."[3]

Thus informal systems have found ways to accommodate people who are not yet part of the formal system, often illegally in structures that have been dissected, segmented and repurposed to accommodate much higher populations than they, and the surrounding infrastructure, were originally designed for.

Such is the case in Hillbrow where, due to the lack of investment interest and falling property values, properties have deteriorated and many have been abandoned by absent owners/landlords, or have been hijacked by dissatisfied or "illegal" tenants who impose themselves as informal body corporates, charging occupants rent but undertaking no management or upkeep of the buildings. The result is that, over the last twenty years, the area has been characterised by increased crime and grime; the disrepair of civil amenities; low levels of social services in the neighborhood; and a shortage of public open space.

Recognising this situation, the Johannesburg Housing Company (JHC) initiated the eKhaya Neighbourhood Improvement Programme in Hillbrow in 2004 – the first of its kind in a high density, predominantly residential inner city neighborhood. In the last decade, eKhaya has developed relationships with property owners, building caretakers, City agencies, residents and other neighborhood community organisations.

To date, the initiative has had considerable success at improving the overall quality of the physical environment particularly in terms of cleanliness, security, maintenance, the upgrading of public spaces; and attracting private and public investment to the area. While the interventions are often in terms of basic urban and facilities management, they have made an enormous difference to lives of people living and working in the area.

---

3   Peter Ahmad. Inner city nodes and public transportation networks: Location, linkages and dependencies of the urban poor within Johannesburg. IHS Working Papers, 2010(26): 11,12.

拓展基金会的新楼是修布罗在20世纪70年代后最早建设的新社会基础设施项目之一，起初占地面积仅有50平方米。为满足所需功能，本地工作室提议将该建筑水平面抬高，在已有建筑上进行延伸从而创造出100平方米的地板面积。于是，20世纪70年代未完成建设的社区中心的交错式屋顶成了新楼实现主要功能的平台。

因此，根据受限的基地和设计任务书，新楼的规划出炉。预算非常有限，只有220万兰特，折合15万美元；方案也相对简单。设计师们了解了这些限定因素之后，便开始考虑还有什么可以从概念上实现。

催化法

《为希望而设计——实现再生持续性的路径》[5]一文中，作者普利托利亚大学的克里斯纳·杜·普莱西斯教授（Chrisna du Plessis）和墨尔本大学的多米尼克·海斯博士（Dominique Hes）认为，专业设计师所面临的问题"在于我们为自己的踌躇犹豫和拖延所编造的故事，在于我们所讲述的可持续性，以及它们在何种程度上受到造成这一问题的世界观和价值体系的强烈影响"。

两位作者将注意力聚焦于一个项目背后的道德思想，提出设计必须质疑和突破已确立的机制，并为置身其中的体系（包括环境体系和社会体系）增添价值，而不是减少价值。这个途径所持的观点认为，害怕扰乱体系可能限制一个项目可能产生的积极影响，而如有"行动的勇气"去挑战现有或已获确认的体系，则能释放一个区域或处境所具有的潜能，产生积极的影响。

这个例子就是如此。建筑师们没有把拓展基金会大楼视为一个物体，而是把这个项目视为一个更为宏大的社会与城市演进过程的一部分——这种方式使得探索社会参与、设计、建设的新维度成为了可能。

这个场所一直以来都是修布罗的休闲空间，建筑师们则希望强化这种体验，并开始为其与周围已建成的构筑物和社区制造互动——通过设计大楼本身的外形，也通过扩大干预周围城市空间的小建设。

5　Dominique Hes, Chrisna du Plessis. Designing for Hope — Pathways to Regenerative Sustainability. New York: Routledge, 2015: 13.

The Outreach Foundation

The site for the Outreach Foundation building falls within the eKhaya Neighborhood precinct, and the new building is used by eKhaya and other agencies active in the neighborhood for meetings. In addition to winning a Gauteng Institute for Architecture (GIfA) Award in 2015, the project has also won the BUILT category in the St Gobain Social Gain Awards 2015 for South Africa; and received a Commendation in the South African Steel Awards 2014.

The Lutheran Community Outreach Foundation, as an organisation, offers support to people living in Johannesburg's inner city through skills development and by inspiring creative journeys through arts enrichment programmes. According to the Foundation, "Urban degradation and unemployment continually result in people feeling angry and dissatisfied with their surroundings; and... the youth... seek guidance from peers and adults who do not have their best interests at heart.

"The various projects offer children, youth and adults the opportunity to engage with arts, culture and heritage activities that facilitate communication, participation and community building; and thus facilitate interaction between residents and the youth; as well as offering alternative activities for children during the school holidays. Through team building, building self-worth and exploring identity the Outreach Foundation offers a holistic approach to families and contributes to communication skills within these communities." [4]

Two years before work on this project started, the Foundation was awarded a grant from the National Lottery to further their objectives. Their intention was to create a facility that brought together a number of their outreach programmes including an active dance programme, music programme, computer learning center, marimba band, and a counselling facility – and to open up the spaces within the existing community center on the site for administration.

The original site was given to the Lutheran church in the late 1890s, and still houses the old church. The site was then added to in the 1960s through a donation by the German Consulate for the purpose of housing a community center – a building that was never completed.

The new Outreach Foundation building, one of the first new social infrastructure projects to be built in Hillbrow since the 1970s,, was initially located on a tiny piece of land of just 50m^2.

---

[4]   Outreach Foundation http://outreachfoundation.co.za/index.php/about/vision-mission.

新楼于2015年初完工，包含三个主要功能，分别在不同楼层为基金会的主要项目提供了场所；这些项目由一道"垂直街道"（一段敞开式的楼梯及画廊空间）连接起来。

在底层，从一个小型入口门厅进入，向南是现有的社区会堂（在交错式屋顶露天平台下方），向东是新计算机中心，后者同时具备学习中心和网吧的功能。

主要的空间，二楼为舞蹈工作室专用楼层，舞蹈课程及活动通过一扇12米宽的玻璃落地窗展示给临近的特威斯特街（Twist Street）。三楼（即顶层）为办公室和会议区，供活跃于该地区的机构和利益相关者使用，如eKhaya社区改进计划、Madulammoho住房协会，以及由拓展基金会运作的一个广播站。

"垂直街道"敞开式楼梯是一个光线充足的流动空间，整个空间被透明的聚碳酸酯板材覆盖。这里变成了一个画廊空间，展出当地自发的作品。通过教授和创造刺绣、织物或编织品等手工艺品的发展机会，Boitumelo项目用艺术品培养自我价值意识，这些手工艺品诉说着制作者的故事，表达着他们的需求——这就使参与者逐渐具有能力战胜贫困，因而有助于促进经济独立。

这些作品就展示在光线充足的公共画廊空间里，它们不仅成了反映地区独特差异与活力的名片，也赋予社区成员对这片空间的主人翁意识。

与公共空间结合

项目首席建筑师托马斯·查普曼（Thomas Chapman）指出，目的是要建造一个修布罗的地标，在已有两幢旧时代建筑物且周边建筑结构密集的地点上建造一座显而易见的崭新的建筑。

项目的设想是建一座小型的、抽象的"灯塔"，这个想法通过采用白色铬锰竹节钢筋和透明的波纹聚碳酸酯作为覆盖材料得以实现。这些覆盖材料不仅塑造出了该建筑作为城市该区域新增元素的简单外形，还发挥了灯塔的主题——在白天，建筑表面是反光的白色，与周围的构筑物形成强烈的对比；在晚上，光线透过透明的聚碳酸酯材料倾泻而出。

查普曼认为，所有建筑师首要关心的应是把公共空间作为设计工作的中心。由于预算的每一分钱都花在了建造大楼上，建筑师们后续发起了一次筹款运动，额外从当地的私营地产开发那里募集了50万兰特（折合3.5万美元）用于建造一个公共的屋顶花园和露天圆形剧场，让

In order to accommodate the necessary functions, Local Studio proposed that the building was pulled up a level and extended over the existing building to create a floor plate of 100m². Thus, the staggered rooftop of the unfinished community center built in the 1970s became the platform on which the major functions of the new building are accommodated.

The planning of the new building therefore emerged from the contained site and client brief. The budget was very limited, a mere 2.2 million ZAR or 150,000 USD; and the programmes are relatively simple. Once the architects knew these defining factors, they then looked at what else could be achieved conceptually.

A Catalytic Approach

In "Designing for Hope – Pathways to Regenerative Sustainability"[5], authors Professor Chrisna du Plessis of the University of Pretoria and Dr. Dominique Hes of the University of Melbourne argue that the problem faced by design professionals "lies in the stories we tell ourselves to justify why we dither and procrastinate, and those we tell about sustainability and how strongly they are still being influenced by the worldviews and value systems that created the problem."

The authors draw attention to the ethos behind a project, proposing that design must question and break open established mechanisms, and add value to, rather than removing from, the system (both environmental and social) of which it is part. This approach takes the view that a fear of disrupting the system can limit the potential for a project's positive influence; while having the "courage to act" by challenging existing or established systems can have positive consequences by allowing the potential of an area or situation to be opened up.

Such is the case in this instance. Rather than seeing the Outreach Foundation building as an object, the architects have chosen to view the project as one part of a larger process of social and urban evolution – an approach that allows new dimensions of social engagement, design, and construction to be explored.

---

5 Dominique Hes, Chrisna du Plessis. Designing for Hope – Pathways to Regenerative Sustainability. New York: Routledge, 2015: 13.

大楼的使用者们通过一个不大却有意义的公共空间与周边的都市空间产生联系。

这个露天平台是与其毗连的修布罗剧院的辅助性空间，曾在修布罗市中心高中戏剧节汇演分区停电期间（为防电站超载而实行的强制断电）用于容纳200名儿童。这个场所也被用于教堂礼拜，并在不久前用于举行2015年4月约翰内斯堡和德班排外暴力事件遇难者的守夜活动。

还有一项很小的辅助性城市干预措施，就是在特威斯特街的对面造了一条钢制长凳，它附着于墙上，为行人在上下班途中休息时观看舞蹈工作室的活动提供了座位。

有一个功能未能在这个中心实现，就是基金会的咨询部门。这个部门将被安排在同一地点的另一座小楼内，这座楼将继续采用与拓展中心一样的灯塔隐喻和语言，内设封闭的咨询空间供创伤咨询工作使用，以及透明的楼梯，以便光线穿透整栋大楼。

技术方面的经验教训

查普曼说，他一直都对工业建筑很感兴趣，因为工业建筑工程师常会想到一些令人激动的替代方案，而这些是建筑师在追求外形时常常忽视的。

当这位建筑师被告知，基于国家彩票的拨款要求，设计和建造这座楼的时间只有6个月时，这种工业灵感就派上用场了。楼的结构设想原本为一个龙门构架，但成本太高。之后有一家供应商找到该建筑师，并说服他采用一种轻量型钢架建造大楼，使用一些结构钢材元素实现更大的跨度。于是大楼最终的建造成本缩减到了每平方米5000兰特/340美元。

由于建筑师们是首次接触这项技术，他们在建设过程中进行了大量的学习。由此产生的一个遗憾就是隔音效果不如预期的那么好，二楼舞蹈工作室的声音会传至楼下的计算机中心。

于是本地工作室又开始筹款，将舞蹈工作室的木地板抬高，与下方的钢结构分离。某种程度上，这对建筑师们是有利的，因为这道工序的额外费用，原预算也是无法支持的。目前情况下，陆续有听说这座楼的人们找到公司，提出帮助寻找解决问题的方案的意愿。

这栋建筑以高度节约资源的方式成功做到了一点，即提供充足的自然光和被动式通风。这是由于舞蹈工作室的三面窗户以及一个垂直的热烟风筒，烟风筒向上穿过了工作室和顶层的

The site has always been a relief space in Hillbrow, and the architectural vision was to enhance this experience and to begin to create interactions with the surrounding built fabric and community – through both the physical form of the building, and by extending small built interventions into the surrounding urban space.

Completed in early 2015, the new building houses three primary functions which accommodate key Foundation programmes on separate levels; and which are connected by a "vertical street" in the form of an open staircase and gallery space.

At Ground Floor Level, a small entrance foyer leads onto the existing community hall to the south (below the staggered roof terrace) and the new computer center to the east, which functions both as a learning center and internet café.

The main space, a Dance Studio takes up the entire first floor; and dance classes and activities are presented to the neighboring Twist Street through a 12m picture window. The second (top) floor accommodates offices and meeting spaces, which are made available to agencies and stakeholders active in the area such as the eKhaya Neighbourhood Improvement Programme and the Madulammoho Housing Association, as well as a radio station run by the Outreach Foundation.

The "vertical street" open staircase is a light-filled circulation space entirely clad in translucent polycarbonate sheeting. This has become a gallery space where the works of a local initiative are exhibited. The Boitumelo Project uses the arts to nurture self-worth by teaching and creating opportunities for handmade product development in the form of embroidered, woven or knitted works that tell the stories and express the needs of the people who make them – thus helping to facilitate economic independence by enabling participants to combat poverty in an incremental way.

These products are displayed in the light-filled, public gallery space which both takes on an identity that reflects the unique contrast and vibrancy of the area, and gives community members a sense of ownership of the space.

办公室。这个系统使得楼内不需要任何机械通风，并且楼内所有的空间在白天都不需要人工照明。

体面的干预

最终建成的建筑不大，却是复杂的城市环境中体面的嵌入；它以最简约的形式满足了任务简介中的基本要求，绝对没有一处多余，却给人明亮、宽敞的体验。

这个新项目的开发也使这个地点整体具有一种前卫感，散发着对未来的希望，预示着新的开始。那具有历史意义的教堂建筑固然是一个庄严、温暖的存在，但久未完工的社区中心却高声诉说着修布罗在过去30年里一直遭受的忽视与匮乏。

通过扩大现有社区中心建筑屋顶的面积，设法在楼内（垂直街道和画廊）和楼外（圆形剧场）都提供有用的、吸引人的公共空间，贴心地在道路对面为路人设置长凳供其观看舞蹈工作室内部——建筑师们面临这项艰巨任务（现存的工地情况、描述的任务简介以及有限的资金），肩负起了个人责任，摒弃退化的城市系统，努力设计了一个具有再生性、适应性、促进性的项目，让修布罗的居民们有了迎接新的社会、文化、经济机会的可能性。

新楼是一个落脚点，一个产生纽带和归属感的地方；这个城市新到来者开始的地方；进行沟通与社区互动的地方。在周边颇具分量的现代主义历史和全球历史氛围中，用当代的语言和全世界通用的技术，一个独特的建筑设想在一个小规模的建筑上得以实现，满足了修布罗社区当地特定的需求以及感情和思想表达的需求。

前行：索菲亚镇的Motswako（"混合"）项目

拓展基金会大楼已被证明不仅是宏大的城市再生进程的一个催化因子，也引发了关于再生性发展的更广泛的公众讨论，其影响已超越建筑领域，延伸至其他行业相关的部门以及公众领域。该项目因其所获奖项众多，并且位于复杂的社会脉络之中，近期吸引了大量媒体关注，也为建筑师们提供了一个重要的机会，为其他专注于再生进程的项目据理力争。

Engaging with Public Space

Thomas Chapman, lead architect on the project, notes that the intention was to create a landmark in Hillbrow, something that was very clearly a new building on a site that already houses two structures from previous eras within a dense surrounding built fabric.

A small, abstract "lighthouse" was envisioned, an idea that was realised through the use of white Chromadek corrugated steel and clear corrugated polycarbonate as cladding materials. In addition to expressing the simple form of the building as a new addition to this part of the city, these cladding materials play on the theme of lighthouse by creating a reflective white surface during the day, which contrasts strongly with the surrounding built fabric; and by allowing light to pour out through the clear polycarbonate at night.

Chapman believes that the primary concern of all architects should be to place public space at the forefront of any design process. Because every cent of the budget was used in the construction of the building, the architects went on a fundraising drive to raise an additional 500,000 ZAR (or 35,000 USD) from private property developers in the area for the construction of a public roof garden and amphitheatre that allows users of the building to connect with the surrounding city through a modest but meaningful public space.

This terrace is a satellite space for the adjacent Hillbrow Theatre and has been used to accommodate 200 children during performances of the Hillbrow Inner City High Schools Drama Festival during load shedding (mandatory power outages imposed in order to avoid excessive load on the generating plant). The space has also been used for church services and, recently, a vigil for victims of the xenophobic violence which took place in Johannesburg and Durban in April 2015.

A tiny, satellite urban intervention also included the construction of a steel bench, which is attached to the wall on the opposite side of Twist Street; and which provides a seating spot for pedestrians to view of the activities in the Dance Studio while resting on their commute.

One of the functions that could not fit into the center was the Counselling arm of the Foundation. This will be housed in a small second building on the site, which will continue the same lighthouse metaphor and language as the Outreach Centre, with enclosed counselling spaces to accommodate trauma counselling work; and transparent staircases that allow light to flood into and out of the building.

拓展基金会新楼完工之后，另一个项目的建设紧接着开始了，那就是约翰内斯堡索菲亚镇 Motswako（"混合"）中心的特雷弗赫德尔斯顿纪念中心（Trevor Huddleston Memorial Centre）。

历史上，索菲亚镇是政治和文化活跃度最高的社区之一，在20世纪四五十年代被黑人占据。受1954年《原住民重新安置法案》（Native Resettlement Act）第19条限制，1955年至1960年间，居民被迫迁往索韦托的梅多兰兹，期间失去了家园，也失去了财产。

旧索菲亚镇少数存留的遗产建筑之一——A.B.叙马宅（A.B. Xuma House），原属于1940年当选的非洲国民大会总统阿尔弗雷德·毕提尼·叙马博士（Dr. Alfred Bitini Xuma）所有，现为博物馆，而Motswako中心与其毗邻。

这座综合建筑于2015年6月开放，白天为索菲亚镇绿色孵化器（Sophiatown Green Incubator）提供场所，该孵化器倡议让来自这个具有历史意义的西部地区社区的年轻人接触更多的创业机会，包括培训和企业孵化；到了晚上，中心变身为一个汇集电影、剧院、舞蹈、艺术功能的演出场地，重拾这片区域昔日的文化氛围。

新楼是对索菲亚镇传统建筑类型学的当代解读，关键是建筑元素如何创造公共开放空间并与之互动。这是通过两大主要设计实现的。索菲亚镇原来房屋的"门廊"（屋前游廊）元素被加以利用，沿着建筑北边的立面创造出一条大型"门廊"，通往大楼的培训和表演空间。这是为了盘活大楼临街的一边——这是历史上本该实现却并未做到的，并为大楼使用者和当地社区成员提供一个可用的户外公共空间。

现有的A.B.叙马宅与新建的中心也界定出一个多用途的庭院，呼应索菲亚镇在20世纪40年代的庭院类型。这片空间属于Motswako中心，但没有设围墙，随时对公众开放。

与拓展基金会一样，这个项目激活了公共空间，有助于增进社会凝聚力与互动；此外，项目对于建筑发扬了过程导向的精神，而非采用以结果为导向的途径。

同样的情况再次出现，由于预算紧张，大楼最重要的元素——覆盖大楼背面和东面的屏幕无法完成。这块金属饰边的屏幕被设计成在功能上可遮光和保护私密性，并将支撑起一幅索菲亚镇的抽象地图，描绘老镇历史上的每一块街区——本质上是一个内置的、露天的交互式展示。借助上面的小金属块纪念物，以前在这里居住过的居民将能找到并标记他们原来的住址。

为实现这一元素，本地工作室在已提供建筑服务并完成部分项目的情况下，积极地进行了一次筹款活动并获得成功。

另外，该项目目标是获得南非绿色建筑委员会（GBCSA）颁发的南非公共及教育建筑五星

Lessons in Technology

Chapman has long been interested in industrial architecture where, he says, engineers often come up with exciting alternative solutions that architects tend to overlook in their quest for form.

This industrial inspiration came in useful when the architect was told that, based on the funding requirements of the National Lottery, he had 6 months in which to design and build the building. Originally the structure was envisaged as a portal frame but the cost was too high. The architect was then approached by a supplier who convinced him to do the building in a light gauge steel frame, with some structural steel elements for the larger spans. This meant that the building was eventually built for just 5,000 ZAR/m^2 or 340 USD/m^2.

Because the technology was new to the architects, they underwent quite a significant learning curve as construction went along. An unfortunate consequence of this was that the acoustic performance is not as good as anticipated in that the sound from the dance studio on the first floor is transferred to the computer center on the floor below.

So Local Studio is now on a drive to find additional funding to raise the timber floor of the dance studio and separate it from the steel structure below. In a way, this has worked in the architects' favour as the additional cost for this process could not have been accommodated in the original budget. As things stand, the firm is now being approached by people who have heard about the building, who are offering to help find solutions to the problem.

Something that the building does do successfully in a very resource efficient manner is to provide ample natural lighting and passive ventilation. This happens through windows on three sides of the dance studio and a vertical thermal chimney that reaches up through the studio, past the offices on the top floor. This system means that no mechanical ventilation has been used at all, and no spaces in the building need artificial lighting during the day.

A Dignified Intervention

The end result is a building that is a modest yet dignified insertion into a complex urban envi-

级绿星证书（5-Star Green Star SA Public and Education Buildings）。为确保成功获得认证，并使项目受到尽可能多的媒体关注，绿色建筑咨询公司（Solid Green Consulting）志愿负责该项目南非绿星认证的工作。

绿色建筑体系包括对先前已开发过的空间进行再利用、设计多功能空间、LED照明及通过太阳能光伏面板供电的节能办公设备、将屋顶雨水收集起来用于灌溉、废弃物源头循环、节水家具、挥发性有机化合物含量低的抛光剂的采用、被动式自然通风（不使用机械通风）、自行车设施，以及能源与水的分项计量。

结论

很多时候，像拓展基金会和Motswako中心这样时间紧、预算少的社会项目几乎没有多少建筑价值，并且基本没有特色。而在这里，建筑师们证明在努力争取的情况下，将项目的紧急性看作一个机遇，还是有可能建成令人神往的建筑物的。

为对不同的历史文脉分别做出恰当的回应，通过积极地使客户与当地利益相关者看到项目的再生性前景，有意识地在建筑元素和公共领域之间制造契合点，这些项目超越了可持续设计，达到了再生性发展的层次。

将它作为更为宏大的城市、经济、社会、环境进程中的设计过程来看，项目结果是一座虽然面积不大也比较简单，却不失体面的建筑。综合这些，可以想象它将成为未来城市投资和积极变化的催化剂。

ronment; and which meets the basic requirements of the brief in the simplest of forms, with absolutely nothing that is superfluous – yet the experience of the building is one of light and spaciousness.

The new development also gives the site, as a whole, a sense of forward movement, of hope for the future, and of a new beginning. While the historic church building is a dignified, warm presence, the unfinished community center spoke volumes, through its incompleteness, of the neglect and deprivation that has characterised Hillbrow for the last thirty years.

In challenging the existing site and brief as presented at face-value – by extending the building's footprint over the roof of the existing community center, by finding ways of providing useful and attractive public space both inside the building (vertical street and gallery) and outside (amphitheatre), and by extending small kindnesses to passers-by in the form of a bench across the road and a view into the dance studio – the architects have taken personal responsibility to move away from a degenerative urban system towards designing a regenerative, adaptive, catalytic project that has the potential to allow social, cultural and economic opportunities to emerge for the residents of Hillbrow.

The new building is a place of arrival, connection and belonging; a threshold for newcomers to the city; a place of communication and community interaction. Within the weight of the surrounding Modernist history, a global history, a particular architectural vision has been realised in a small-scale structure with a contemporary language, using technology that can be found anywhere in the world, to accommodate the very particular needs and expressions of the local Hillbrow community.

Moving Forward: The Mix in Sophiatown

The Outreach Foundation building has proven to be not only a catalytic element in a larger urban regeneration process, but has also contributed to a broader public discourse on regenerative development that is reaching beyond the profession of architecture into other industry-related sectors and the public. The project has recently attracted considerable media attention, both through the awards that it has received and because of its location in a complex social context – and has provided an important opportunity for the architects to argue for other projects that focus on the process of regeneration. Immediately after the Outreach Foundation was completed, construction started on the Trevor Huddleston CR Memorial Building at the Mix in Sophiatown, Johannesburg.

Sophiatown was historically one of the most politically and culturally active neighborhoods occupied by black people in the 1940s and 1950s. Under the Native Resettlement Act No. 19

of 1954, residents were forcefully removed to Meadowlands in Soweto between 1955 and 1960, losing both their homes and possessions in the process.

The Mix is situated adjacent to one of the few remaining heritage structures from the old Sophiatown, the A.B. Xuma House, which belonged to Dr. Alfred Bitini Xuma who was elected president of the African National Congress in 1940, and which is now a museum.

Opened in June 2015, the building complex accommodates the Sophiatown Green Incubator by day, an initiative that connects young people in the historic western area neighborhoods to entrepreneurship opportunities including training and enterprise incubation; and by night, the center harks back to the area's cultural past by transforming into a performance venue hosting film, theatre, dance and arts ventures.

The new building is a contemporary interpretation of traditional Sophiatown building typologies – including, importantly, how the built elements create and interact with public open space. This happens through two primary devices. The traditional "stoep" (verandah) element of the original Sophiatown houses has been used to create a large "stoep" along the northern façade of the building, onto which the training and performance spaces of the building open. This serves to activate the street edge of the building as it would have been historically, and provides an outdoor public space for use by building users and members of the local community.

The existing A.B. Xuma House and the new building also define a multipurpose courtyard, echoing the yard typologies of 1940s Sophiatown – a space that has no boundary walls and is open to the public at all times.

Like the Outreach Foundation, in addition to activating public spaces that contribute towards improved social cohesion and interaction, this project engages the ethos of a process-oriented, rather than an object-oriented approach to architecture.

Once again, the tight budget prevented the completion of one of the building's most important elements – a screen that wraps around the north and east façades. Designed to function as a shading and privacy device, the metal lace screen will support an abstracted map of Sophiatown to depict every block of the historic old neighbourhood – essentially a built-in, open-air, interactive exhibit. Former residents will be able to find and mark the locations of their original homes with remembrance plaques on this map.

To make this element happen, Local Studio has actively, and successfully, undertaken a fundraising drive over and above the architectural services that the practice has provided.

In addition, the project is targeting a 5-Star Green Star SA Public and Education Buildings (PEB) certification from the Green Building Council of South Africa (GBCSA). In order to ensure that this process is successful and that the project receives as much media attention as possible, green building consultants Solid Green Consulting have undertaken the Green Star SA certification process on a pro bono basis.

Green building systems include reuse of a previously developed site; design of multi-functional spaces; LED lighting and energy efficient office equipment run off solar photovoltaic panels; rainwater harvesting from the roof for irrigation; waste separation at source; water efficient fittings; low VOC (volatile organic compound) finishes; passive natural ventilation (no mechanical ventilation used); cyclist facilities; and energy and water sub-metering.

Conclusion

All too often, social projects on tight timelines and tight briefs like the Outreach Foundation and the Mix have very little architectural merit and are quite nondescript. Here, the architects have proven that it is possible, under trying circumstances, to view a project not as an emergency but as an opportunity, and to make the building a desirable place to be.

In responding appropriately to their respective historic contexts, by actively engaging clients and local stakeholders in the regenerative visions of these projects, and by creating intentional dovetailing between the built elements and public realm, these projects have been taken beyond sustainable design into the realm of regenerative development.

Viewed as design processes within larger urban, economic, social and environmental processes, the result is an architecture that, while modest and simple, is dignified. And when these things come together, it is possible to imagine a catalyst for future urban investment and positive change.

当代拉丁美洲建筑：
常识与理想主义

[巴西]露斯·沃得·采恩
译者：伍雨禾（英译中）
校译：张晓春

拉丁美洲[1]，包含三十多个国家，具有丰富多元的文化、政治与地理景观。因此，想通过一个建筑来代表拉丁美洲当代建筑的多样性与复杂性，几乎是个无法完成的使命。而"多样性与复杂性"却正是上海2015年CICA论坛的组织者要求参会者们诠释的主题。当然，从已在主要国际杂志或网站上露面的那些眼熟的当代拉丁美洲建筑中找到一个例子并非难事，然而，笔者倾向于选择一个并不为人熟知、规模小且低调的建筑。笔者相信位于巴西圣保罗的"Projeto Viver"住宅项目将是一个完美的选择。通过这个案例，还可以引入一些既有趣又有意义的当代议题，这些辩论一直渗透在拉美乃至全球不断变化的城市与建筑的现状之中。

2014年，"Projeto Viver"住宅项目赢得了一个拉丁美洲相当重要的建筑大奖，"罗杰里奥·萨尔莫纳拉美建筑奖：开放空间/集体空间"（Rogelio Salmona Latin American Architecture Award：Open spaces/Collective spaces），该奖项创立于2013年，由总部位于哥伦比亚波哥大的罗杰里奥·萨尔莫纳基金会组织与赞助。尽管目前拉美之外的地区对这一奖项的了解十分有限，但其重要性与认可度在未来的十年中将呈指数增长。

在描述该奖项如何创立以及如何评奖之前，有必要先介绍一下罗杰里奥·萨尔莫纳（Rogelio Salmona，1929—2007）。他出生于法国，很小的时候就跟随家庭移民至哥伦比亚。萨尔莫纳出色的绘画技能与超乎常人的天赋很早就被发掘了，当他还是建筑系学生的时候就被指派给柯布西耶做其波哥大规划的助手。1948年他搬到巴黎并加入了柯布西耶的"Rue de Sèvres"工作室，之后在法国住了十年并周游了欧洲。1958年萨尔莫纳回到哥伦比亚后，他

---

[1] 巴西的国土面积约为851.6万平方公里，大约比中国的国土面积（约960万平方公里）小10%。拉丁美洲的面积（包括加勒比地区）约为2107万平方公里，超过中国和巴西两国面积之和。

Contemporary Architecture in Latin America: Common Sense and Idealism

Ruth Verde Zein

Latin America is almost a continent[1], encompassing three dozens of countries and a varied assortment of cultural, political and geographical realities. As so, it is a very difficult task to find just one building apt to represent the diversity and complexity of its contemporary architecture. Yet, that was exactly the briefing presented by the organizers to the participants of CICA Shanghai 2015 Seminar. Certainly, to represent our subcontinent, it would be easy to choose one building among several other slightly familiar contemporary buildings that had already been published in one of the main international magazines and websites. Instead, I have preferred to select a not so well known, quite small and low-key piece of architecture. I believe that was a good choice, and that this building – the Projeto Viver headquarters in São Paulo, Brazil – is a perfect case. Among other reasons, because it allows me to introduce some meaningful and interesting contemporary debates, that are constantly permeating the ever-changing conditions of our urban and architectural realities in Latin America - and perhaps, elsewhere.

Projeto Viver building won in 2014 a very important Latin American Architecture Award. It is a Prize that has been created in 2013, and it is still little known outside the continent, but its importance and acceptance will exponentially increase in the next decade. It is organized and granted by the Rogelio Salmona Foundation, whose headquarters are situated in Bogotá, Colombia; and its name is "Rogelio Salmona Latin American Architecture Award: Open spaces / Collective spaces".

---

[1] My country, Brazil, has 8,516,000 km², a around 10% smaller than China (about 9,600,000km²). Latin America area (inclusive the Caribbean region) exceeds that of both together (21,070,000 km²).

的作品逐渐展现出一种个人对建筑的批判性思考，同时也渗透着对地方需求、地域特征与城市文脉的关注。他的态度与同时代的其他拉美建筑师发生了共鸣，比如乌拉圭的艾拉迪欧·迪斯特（Eladio Dieste）和墨西哥的卡洛斯·米哈雷斯（Carlos Mijares）。一些有关拉美建筑的史学观点认为，他们代表了"另一种"现代主义建筑的趋势、一种"适度的"现代性。[2] 萨尔莫纳坚定地认为应当视建筑设计为改善城市的质量与宜居性的机会，他的作品总是力争创造更多开放的公共空间，这些空间为使城市变得更友好、更民主发挥了积极的作用。后来，世界范围内的评论家与史学家都将萨尔莫纳看作是拉美第三代现代建筑师中最重要的成员之一。他去世之后，为了纪念他的杰出贡献，在哥伦比亚波哥大成立了罗杰里奥·萨尔莫纳基金会。[3] 为了传承这份记忆，基金会在2013年设立了"罗杰里奥·萨尔莫纳拉美建筑奖：开放空间/集体空间"。这个两年一度的大奖在2014年举办了第一届，第二届在2016年完成。大奖的使命就是发现、认可并鼓励拉丁美洲与加勒比城市的优秀建筑实践，嘉奖那些高质量的、为城市与市民提供了重要而愉快的集体开放空间的建筑，由此为发展和巩固更具包容性的都市空间作出贡献。

这个奖项的独特之处在于，它只授予那些到评奖之时已使用五年以上并仍然处于良好状态的建筑，只有使用中的建筑才能证明其空间是否能够鼓励市民自发地聚集。不论是公共建筑还是私人建筑，奖项看重的是在设计质量与包容性上的突出表现。只要设计概念力图创造具有持久性、激动人心的集体空间，任何功能、尺度与规模的建筑都能有机会获得提名。简而言之，此奖的目的在于嘉奖那些用谦逊的方式改进了各自所处城市的环境、发掘并提升了所处场所的氛围以及附近城市生活的价值的建筑。

尽管任何人都可以在基金会的网站上提交建议，但萨尔莫纳奖并不是一个公开竞赛。一个国际专家委员会经过资料梳理与慎重考虑后选出候选建筑，此委员会包括四名成员及其各自身后的团队，他们分别代表了拉美四个主要区域。2014年CICA的合作发起人、墨西哥艺术与建筑评论家路易丝·诺勒（Louise Noelle）是中美洲和加勒比地区的代表；SAL（拉美论坛）的发起者之一、哥伦比亚建筑史学家与评论家西尔维亚·阿朗戈（Silvia Arango）代表了南美洲安第斯山脉国家；Summa+杂志编辑、建筑评论家、霍尔希姆奖（Holcim Awards）国际评审团成员费尔南多·迪耶兹（Fernando Diez）则作为南美洲南方腹地国家的代表；笔者负责了第四个地区——即巴西地区[4]，这四名成员及其团队负责各自地区建筑作品的遴选。

首届评奖期间，国际组委会专家们在与各自团队的共同努力下，仔细思考、研究并参观了超

2　Enrique Browne. Otra Arquitectura en America Latina. Mexico, Gustavo Gili, 1988. Marina Waisman, Cesar Naselli. 10 Arquitectos Latinoamericanos. Sevilha: Dirección General de Arquitectura Y Vivienda, 1989. Antonio Toca. Nueva Arquitectura en America Latina: presente y Futuro. Mexico: Gustavo Gili, 1990. Jorge Francisco Liernur. America Latina. Architettura, gli ultimi vent' anni. Milano: Electa Editrice, 1990. Cristian Fernandez Cox, Antonio Toca Fernandez. America Latina: nueva arquitectura. Un a modernidad posracionalista. Mexico: Gustavo Gili, 1998. Hugo Segawa. Arquitectura latinoamericana contenporánea. Mexico: Gustavo Gili, 2005.

3　Fundación Rogelio Salmona, http://obra.fundacionrogeliosalmona.org/ Latin American Award, http://premio.fundacionrogeliosalmona.org/

4　笔者仍作为2016年萨尔莫纳奖巴西地区的评委代表。

Before describing how this award was created, and how the competition to choose the winner is being organized, it is necessary to explain who was Rogelio Salmona (1929-2007). Born in France, his family migrated to Colombia when he was very young. His excellent drawing skills and superior talent were early recognized: he was appointed to work with Le Corbusier for his Plan to Bogotá when he was still a student of architecture. In 1948 he moved to Paris and joined Le Corbusier's Rue de Sèvres Atelier. He lived in France for a decade, having travelled all through Europe and beyond. When he returned to Colombia in 1958 his works revealed a personal critical approach to architecture, permeated by a more attentive regard to the local necessities and resources of each region and urban context. His attitude was in resonance with that of other Latin American creators of his generation, as for example, Eladio Dieste in Uruguay or Carlos Mijares in Mexico, all of them considered by some historiographical revisions on Latin American architecture as representing "another" modern trend, the "appropriate" modernity[2]. Salmona strongly believed that the design of any piece of architecture had to be treated as an opportunity to enrich the quality and livability of the city. His works always strived to include public Open spaces that positively contributed to a more friendly and democratic city. Finally, world-widely critics and historians recognize Salmona as one of the most important Latin America's third generation modern architects. After his demise, the Rogelio Salmona Foundation[3], established in Bogotá, Colombia, is preserving his legacy. They launched in 2013 the Rogelio Salmona Latin American Architecture Award: Open spaces / Collective spaces, to further celebrate his memory. The first edition of the biannual Award happened in 2014, and the second edition was launched and will be completed in 2016. The Prize mission is to identify, recognize and stimulate the dissemination of good architectural practices in Latin American and Caribbean cities. It seeks to put in value high-quality buildings that also generate significant and convivial open and Collective spaces for the city and its inhabitants, thus contributing to the development and consolidation of more inclusive urban spaces.

A unique characteristic of this Award is that it is granted only for buildings, which had already had at least five years of being used, and is still in proper and good use at the moment of their selection for the competition. The Award is granted either to public or to private buildings,

---

2   Enrique Browne. Otra Arquitectura en America Latina. Mexico, Gustavo Gili, 1988. Marina Waisman, Cesar Naselli. 10 Arquitectos Latinoamericanos. Sevilha: Dirección General de Arquitectura Y Vivienda, 1989. Antonio Toca. Nueva Arquitectura en America Latina: presente y Futuro. Mexico: Gustavo Gili, 1990. Jorge Francisco Liernur. America Latina. Architettura, gli ultimi vent'anni. Milano: Electa Editrice, 1990. Cristian Fernandez Cox, Antonio Toca Fernandez. America Latina: nueva arquitectura. Un a modernidad posracionalista. Mexico: Gustavo Gili, 1998. Hugo Segawa. Arquitectura latinoamericana contenporánea. Mexico: Gustavo Gili, 2005.

3   Fundación Rogelio Salmona, http://obra.fundacionrogeliosalmona.org/ Latin American Award, http://premio.fundacionrogeliosalmona.org/

4   For the 2016 edition of the Salmona Prize some names changed, but luckily, I was confirmed as Brazil's representative.

过100个建筑（均建于2000—2008年），并从中选出了21个入围2014年萨尔莫纳奖的候选作品，它们分别来自阿根廷、智利、哥伦比亚、玻利维亚、巴西、厄瓜多尔、墨西哥、秘鲁、波多黎各与委内瑞拉。基金会联系了这些建筑师，愿意参与的评奖者寄来图像资料，基金会编辑了图书专辑并组织了一次展览。最终的评委会除了上述四位国际委员会专家之外，还加入了第五位评委，以此来增加国际视野。2014年萨尔莫纳奖的特邀评委是日本建筑师内藤广（Hiroshi Naito）。重新审查所有候选项目之后，经过三天的激烈讨论，评审会选定了最终大奖以及三个荣誉提名奖。

寻找符合奖项要求的——设计精良、符合语境且维护良好的拉美建筑是一项很大的挑战，但出乎意料的是，评选并非如最初预料的那样难。为了体现21世纪的时代性，第一届大奖的参选建筑均建于2000—2008年，2016年奖项面向建于2008—2011年的建筑，之后则依此类推。或许现在这样说还为时过早，但可供选择的案例逐年增多确是不争的事实，这使我们的工作变得越来越容易。如今在拉美年轻并富有经验的建筑师中好像正涌起这样一股势头：把为21世纪的城市创造更多有价值的公共空间当作目标。也许这只是一个愿望，但我们期待这个乐观预言成真：我们的城市"气候"将在未来的数十年中越变越好，尽管危机和困境仍困扰着拉美大陆——或许危机也意味着机遇。

"塑造城市与公共空间的建筑"是对于大奖宗旨的巧妙概述，在这个很简单的理念中包含了极为多样的情状与形态，国际评委会成员发现主要的难题是如何更好地理解并整理有价值的案例。有一点很明确，我们面对的并不是传统的、19世纪基于欧洲理念的城市。我们试图理解、共处甚至赞美的是当代拉美城市，不应背负那些概念过时、如幽灵般存在甚至妨碍我们恰当理解现实的沉重负担。拉丁美洲新的城市建筑内容丰富、形色各异。不同规模与风格的建筑尖锐并置，反差强烈，形式与经济因素并不调和，表现出随意性（非正规性）与不连续性的特点。这些城市建筑，一方面面临着巨大的环境与社会问题，另一方面受到强大的投机与金融力量的操控。这样错综复杂的状态构成了拉丁美洲当代建筑的一种不稳定的整体特征，使人很难对其形成理性的理解，讨论和寻找解决问题的途径就更为艰难了。因此，若是以草率或偏见的眼光来看待拉美城市，我们将震惊地看到由无趣的、丑陋的以及反乌托邦的物体组成的一团乱麻，建筑物之间被碎片化的、结构松散的空间分隔开。如果事实如此，那么我们是否还能找到以"塑造城市与公共空间的建筑"为宗旨的优秀当代建筑？

答案是肯定的。在巴西乃至拉丁美洲，我们的确找到了相对而言较少（相较我们城市的巨大尺度而言），但仍然数量可观的一些特殊案例，它们都值得进行仔细研究并被纳入萨尔莫纳奖的考虑范围。它们中的一些项目经由政府业主的努力而建成，一些项目由非政府机构业主推动建成，也有几个源于私人机构业主的努力。其中有的建筑已经在世界范围内小有名气，其余的建筑虽然名气不大，但却是低调而有趣的，它们往往就像在一大锅常识中放一小撮理想主义进行调味。这些案例，有的嵌入著名公园或大型公共空间中；有的建于市中心高密度区块或居住区中；有的建于大都市扩张形成的不稳定的外围区域。在资金方面，有的预算充

provided that they stand out for the quality and generosity of their design. Buildings of any program, dimension or scale, may be nominated to the Award, as long as their design conception had strived to include lasting and stimulant open Collective spaces. The nominated buildings should be in use, thus giving evidence that their architectural spaces had actually been able to encourage the spontaneous gathering of citizens. In brief, the Award purpose is to celebrate buildings that are collaborating to the betterment of their respective urban environments, and doing so in a respectful way: by recognizing and valorizing its place, the surrounding architectural ambiance and the nearby urban life.

The Salmona Prize is not an open competition, although any person may submit suggestions at the Foundation's website. The buildings are selected, after the careful consideration of all information available, by an International Committee of experts, consisting of four members, representing Latin America's four main regions, helped by their respective local teams. In 2014, the CICA co-founder, Mexican art and architecture critic Louise Noelle represented Central America and Caribbean region. Colombian architectural historian and critic Silvia Arango, one of the founders of the SAL (Latin American Seminars), represented South America Andean countries. Editor of *Summa+* magazine and architectural critic Fernando Diez, also a member of the Holcim Awards International Jury, represented deep South American countries. And I was in charge of the selection of works of the fourth so-called region, my country, Brazil[4].

In the first edition of the prize, due to the combined efforts of the International Committee experts and respective local teams, more than a hundred buildings (built between 2000 and 2008) were carefully considered, studied and visited. Twenty-one buildings from Argentina, Chile, Colombia, Bolivia, Brazil, Ecuador, Mexico, Peru, Puerto Rico and Venezuela were selected for the 2014 Salmona Award. The authors were contacted by the Foundation, and those who had agreed to participate sent graphic material to organize an exhibition and a book. The final jury included the four already mentioned international committee experts plus a fifth juror, in order to give us a much needed foreign perspective. In the 2014 Salmona Award the guest juror was the Japanese architect Hiroshi Naito. After the careful reexamination of all candidates, and three days of intense debates, the jury decided to grant the Prize, plus 3 Honorable Mentions.

It was a challenge to try and find well designed, contextually meaningful and well maintained Latin American buildings able to fulfill the Award's requirements. But surprisingly, it was not

---

4   For the 2016 edition of the Salmona Prize some names will change, but luckily, I was confirmed as Brazil's representative.

足，有的则捉襟见肘。这些建筑随机地散落在各自所在的城市空间中。无论如何，笔者认为这些备选作品从根本上而言都是非常出色的建筑——它们展现出的美学品质完全符合评委的选择标准。

获奖建筑出自拉丁美洲不同年代、不同地域的不同建筑师之手，但是，建筑师们都在为同一种可能而努力：城市将因他们在设计中小心营造的那些开放空间而发生好的变化，即使这只能在相当小的规模上影响城市。这些作品以及建筑师的讨论表现出，他们相信能够"塑造城市"的优秀建筑或许有能力潜移默化地引导城市转向更好的发展。即使转变的方式不那么引人注意，这些案例仍燃烧着古老的、几已熄灭的乌托邦火焰。笔者在此提到"乌托邦"这一过时的概念并非把它当作一套具有魔力的观念，能瞬间彻底改变建筑、城市与居民，而是认为"乌托邦"意味着一个更简单更适度的概念；或者用它可以更好地描述一种具有挑战性情景的特征：即一种相信仍然有机会将世界变得更好的信念；而建筑必将在这种变化中发出自己的声音。这也是最初打动罗杰里奥·萨尔莫纳奖组织者们的信念。

"Projeto Viver"住宅项目

经过评委会的精彩辩论，在相互提问并学习他人观点后，最终评委会成员认为第一届萨尔莫纳奖不应颁给一个大型建筑，而应当授予一个非常谦逊而独立的小建筑。FGMF事务所在圣保罗设计的"Projeto Viver"项目赢得了评委的青睐。

若是仔细探讨2014年萨尔莫纳奖的其他20个候选项目，将会很有趣，对其他候选项目大致特征的了解为更充分地理解"Projeto Viver"项目提供了一个关于拉美当代建筑的全景认知，能够帮助笔者更好地介绍获奖作品。然而因篇幅有限，笔者这里只能介绍"Projeto Viver"项目。同时，笔者并不希望让读者将"Projeto Viver"项目看作一个孤立的优秀建筑。要理解这座建筑，除去审视作品本身，也应当看得更远：它标志着应当如何看待和对待21世纪的拉美城市。不论怎样，如果该建筑并不是希冀提供一个简单或全方位的方案，那么它至少包含了一个宽泛的命题：在我们半整合状态的城市中，有时我们必须以小见大。为了读懂这座建筑、读懂拉丁美洲的城市——看出它们的本质、看出它们的未来，我们必须驶离过去数个世纪的停滞、那些传自欧洲的观念，并且跳脱开所有那些美丽的偏见。从很多方面而言，这座谦逊的建筑呈现了一个积极而又美好的未来——但是我们须得擦亮眼睛方能充分认识到这一点。

"Projeto Viver"项目坐落于圣保罗市的莫隆比（Morumbi）社区，此区域中豪宅、公寓住宅、社会住宅项目以及贫民窟都挤在一起。"Projeto Viver"项目的目标是安置那些原本生活在雅尔科伦坡（Jardim Colombo）贫民窟的贫困居民。一个名为"居者有其屋"协会（"Viverem

as difficult as we initially thought it would be. In order to catch up with the 21st century, the Award first edition encompassed the buildings inaugurated between 2000 and 2008. The 2016 Award edition will include buildings inaugurated between 2008 and 2011; and so on. It is early to say, but it seems that our task is going to become progressively easier in the years ahead. We have the feeling that there is perhaps a springing trend growing among Latin America's young and experienced architects: an aim to contribute to the production of meaningful public spaces for the 21st century cities. Maybe this is just wishful thinking. But let's hope my optimistic forecast is correct, and our urban "weather" is about to change for the better in the next decades, despite crises and difficulties that strive our subcontinent — or maybe, because of them.

Considering the sheer diversity of situations and modalities that may be included in the apparently simple idea of "architecture that helps make city" — a lead that ably and briefly summarizes the Award's feature and intention, the main difficulty that we, the International Committee members, had found, was how to better understand and sort out the meaningful examples. For us, it was quite clear that we were not dealing with the traditional, 19th century, European-born idea of the city. We tried to understand, to work with, and if possible, to praise, our Latin American cities for what they are, without burdening them with the heavy shadow of anachronistic conceptual ghosts, unfit to aptly understand our realities. As we know, Latin American cities are relatively new constructs of varied dimensions, huge contrasts, stark juxtapositions of scales and styles, formal and economical incongruities, informalities and discontinuities, with vast environmental and social problems on the one hand, and strong speculative and financial powers dominating the floor on the other hand. This complex mixture results in an apparently unbalanced ensemble of very difficult rational apprehension, and even more difficult and concerted solution. So, if we observe our cities with a hasty or biased look, we tend to be shocked, and to perceive them only as an amorphous set of uninteresting, dystopian and ugly objects isolated by obnoxious in-between voids made of debris or unconsolidated fabrics. If that were so, would there be any room for finding good meaningful contemporary buildings to be considered under the lead of "architecture that helps make city"?

Perhaps surprisingly, yes there was. In Brazil, in Latin America, we were actually able to identify a perhaps relatively small (considering the huge size of our cities), but still most significant number of exceptional buildings that deserved to be more carefully examined and considered for the Salmona Award. A few of these buildings were born from important governmental efforts; a number resulted from the initiative of non-governmental agencies, and several from generous initiatives of private agents. Some were buildings that had already been internation-

Família")的非政府组织常组织各种活动,其任务是促进该社区的发展。这块30米×50米的基地是附近区域最后一片可用的空地,曾被用作非法的垃圾堆放场。尽管这块空地卫生条件极差,影响健康,但贫民窟居民们仍要每天穿越它到达街区的内部街道,而且居民们把这里当作公共的活动场所,踢球或举行社区活动。从方案伊始,建筑师即非常尊重场地特质中好的一面,他们力图维持场地的开放性,以便居民仍能自由穿越并展开休闲活动。设计的第一步是彻底重新设计原有机动车道与人行道的入口。

"Projeto Viver"项目的建筑部分沿场地边沿布置,从而将原先崎岖不平、混乱不堪的场地限定出一个多层广场。于是,这块空地的中心并未被建筑占领,而是设计成一个开放的广场,供人们静思、嬉戏,还可当作露天看台观看现场表演。项目还保留了公共道路与贫民窟之间的空隙。

建筑功能被分置在两个建筑体量中。其中,纵向场馆紧贴基地西界,另一个横向场馆与之垂直,底层架空,使人们可以自由穿越场地进入贫民窟。这两个建筑体量限定了公共的开放空间,并且将其组织成两个部分:阶梯形广场入口区域以及底层多功能运动场地。横向场馆底层架空的部分被作为有遮盖的长廊,供人们举办各种或正式或非正式的活动;二层布置多用途教室、会议室、行政办公室以及一个小图书馆;而在柱廊以下的地下空间则设有更衣间、储藏室和公共浴室,这些都由当地社区来组织管理。

纵向场馆的首层设有前台、警卫室以及一个供多学科使用的手工艺车间。建筑的大门是可以完全敞开的,以便在有需要时可以将室内空间向公众开放。二楼布置了一系列小房间,包括医药室、牙医室、法律咨询、心理咨询室以及一个等候室。面向街道还有一间体验厨房为社区提供烹饪课程,此外还设有一个对外销售自制产品的小商店。屋顶平台有一个小的儿童活动室,露台上也会开展一些园艺、瑜伽等课程。总之,空间的功能组织相当灵活,可以根据需求随时调整。在经过几年使用之后的今天,就像预期的那样,一些原有功能和活动根据非政府组织的安排及社区不断变化的需要顺利地进行了转换。

两个建筑体量通过空中走廊相连接,位于二层的房间——尤其是那些供社区使用的部分——都相当开放和通透,表达出欢迎使用的姿态,有效地消融了住区组织刚搬进这总部时给居民带来的不信任感。

建筑结构相当简洁,使用的是钢筋混凝土墙和混凝土砌块墙。一些金属窗框外包有镀锌钢百叶,另一些则饰有金属类百叶构件,以便白天时关闭百叶为房间遮阳。楼梯与走道都由金属网系统构成,包括手工艺车间上层的门和封闭管理室的穿孔板也使用相同的金属网系统。建筑中有意使用了城市贫穷地区常见的简单材料,例如混凝土砌块和碎陶片。但是在这个项目中,这些材料呈现出不同且富有创新意味的使用方式:建筑并未以直接模仿的方式融入这一区域,而是依靠发掘其中已有的元素并对其进行重构,使之产生新的美学意义。

ally well publicized; many others were somewhat discreet but very interesting examples, designed with huge doses of common sense, spiced with a pinch of idealism. Some were buildings inserted in prestigious parks and consolidated public spaces; others were found in dense central areas and neighborhoods; others, in the sprawling and precarious metropolitan peripheries. Some had been built with the benefit of large budgets; others had been erected with a very tight economy of means and resources. All of them were found randomly punctuating their respective cities. In any case, the selected works were also, or may I say, primarily, some very good pieces of architecture — since their esthetical qualities were not at all a secondary aspect of our selection criteria.

Each of these buildings was designed by one out of an ample variety of architects, of different generations, working in many different places of our large subcontinent. Yet, all the authors seemed to have worked on the possibility that their cities would perhaps experience some good changes with the help of the, sometimes very discreet, Open spaces created by their buildings, even if they would affect the city only in a modest small scale. These works and sometimes their authors' discourses seemed to trust that a good example of good architectural design, of an "architecture that makes a city", might be able to have a say, albeit small, in helping to transform and to contaminate their places, for the better. Perhaps, even if in a most inconspicuous way, all theses examples still carry on the old, almost burnt off, flame of utopia. I'm sorry to bring up here again this apparently outworn concept. But here, I do not want it to mean something fixed, like a set of closed ideas carrying a magical power that would suddenly and completely change forever and ever architecture, the cities and their inhabitants. The word utopia, here, is meant as a more simpler and modest concept. Or else, it is here used just to better characterize a challenging condition: the belief that there is still some possibility of changing the world for the better, even if punctually, even if only in our surroundings, even if just in minuscule doses; and that architecture may still have a say in that possible change, too, which is exactly the belief that has primarily moved the organizers of the Rogelio Salmona Prize in the first place.

Projeto Viver

As it happens, I was very surprised when, after one of two days of debates, the members of the Salmona Award Jury gradually drifted into considering the possibility of choosing a Brazilian building as the winner of the 2014 edition of the Prize. Although I considered it as a very fine example of architecture, it was not initially my favorite, but the Chilean examples.

最后，对建筑师做下介绍。建筑师在设计这所建筑时刚从业五年。2000年，年轻建筑师费尔南多·福特（Fernando Forte），劳伦斯（Lourenço Gimenes）和罗德里戈（Rodrigo Marcondes Ferraz）创建了FGMF建筑事务所，之后他们的事务所逐渐发展。15年间，他们已经完成或参与了350个项目。尽管项目数量如此庞大，他们的事务所仍保持着一个相对小型的规模，同时，尽可能多地寻找机会以探索不同规模、材料和需求的建筑实践。项目上的多样性使得他们能够在设计中一直表现出较高的研究精神，并且在管理中善于在产量高与设计精之间保持平衡。他们相信，对于错误的批判性包容是通往好设计的关键所在，或许，正是像这样反形式主义或者"不走寻常路"的方式，帮助他们实现了一种聪明的可适性，使他们善于处理各种不同的复杂情境。

拉丁美洲这片大陆，在我看来，仅用一个建筑来代表它的当代建筑的丰富性，即使只是想想都觉得荒谬。不论如何，希望这个小巧而低调的建筑能够为读者展示一些拉丁美洲城市与建筑的现状——一种多元的状态。

（本文所有图片均由作者提供）

However, during the very interesting jury debates, by learning from each other's opinions and questionings, my position gradually changed. Eventually, the Jury members realized that it was better if the first Salmona Prize edition would not go to a large scale example, but instead, that it should be granted to a very modest and independent example. So we chose the Projeto Viver Building, in São Paulo, designed by FGMF Architects.

It would be very interesting to carefully consider all 20 buildings that were selected for the 2014 Salmona Award; or at a minimum, the 3 Honorable Mentions. But since the space here is limited, I will have to stick only with the Prizewinner. Alas, a careful consideration of the other buildings would better introduce the winner, in the sense that their features would help prepare the terrain to fully understand the qualities of Projeto Viver. If not for other reasons, because I would not like you to look at that building as an isolated piece of good architecture –which it certainly is. Besides looking at this building in itself, it should also be understood beyond that: as a sign of how Latin America 21st century cities should be thought of, and acted upon. In any case, if this building does not contain a simple or all-encompassing formula, it may contain a broad proposition: that in our half-consolidated cities, sometimes we have to think small to think big. In order to look at this building – and to our Latin American cities – for what they are, and for what they could be, we have to sail away from the moorings of previous centuries, European-born ideologies, and to get stripped of all such petty prejudices.

In many ways, this modest building represents a possible and fruitful future – but we have to refresh our eyes to fully realize that.

The Projeto Viver building is situated in the Morumbi neighborhood of São Paulo city, an area of that houses side by side both luxury villas and condominiums and popular housing "projects" and informal slums. Projeto Viver building was designed to attend the underprivileged population from the deprived Jardim Colombo community (favela). It hosts the activities of the non-governmental organization "Viver em Família" (Living in Family Association), whose mission is to help the human development of that community. The 30 x 50 meters lot was the last open area available in the vicinity, and was formerly misused as an illegal garbage disposal place. Despite that insalubrious condition, it was daily crossed by the favela residents in order to access the internal streets of the neighborhood, and precariously used as a common living area for playing soccer and other social events of the community. Right from the very start, the architects of the Projeto Viver building acknowledged these features from their better side, and strived to maintain the place open for the free crossing and the leisure of the neighbors. Their first step to define the parti was to adequately redesign the preexisting accesses for vehicles and pedestrians.

Instead of the existing sloppy and ravenous terrain, the Projeto Viver building and its Open spaces defined a multi-leveled square, saving the gap between the public street and the favela. As so, the center of the terrain is not occupied by the constructions, but to be freely used for resting, an open plaza for contemplation, children's games, and the bleachers used for open air shows.

The functional program itself was split into two pavilions. A longitudinal pavilion is placed next to the western limit of the parcel; a second pavilion is placed transversally, retired in the back, over pilotis, so as not to impair the crossing of people accessing the favela. Both pavilions define the public/Open spaces, and organize them in two sectors: the graded plaza access and the lower flat multi-sports court. The pilotis of the second pavilion may be used as a kind of covered veranda where multiple programmed and/or informal activities may happen. The transversal pavilion houses above multipurpose rooms for classes, meetings and administrative uses and a small library. Underground, bellow the pilotis level, there is a basement with changing rooms and deposits that are used by the players and/or as a communal bathing facility, whose management is organized by the local neighbor's association.

The longitudinal pavilion ground floor houses the reception, the janitor's house and a multi-disciplinary workshop with a large swinging door that may be used to open the premises to

图1—图5:"Projeto Viver"住宅项目;São Paulo, FGMF 建筑事务所,
2014"罗杰里奥·萨尔莫纳拉美建筑奖:开放空间/集体空间"获得者
fig.1-5: Projeto Viver Building, São Paulo, FGMF Architects. Winner of the
2014 Rogelio Salmona Latin American Architectural Award: Open spaces /Collective spaces

the public spaces, when necessary. On the upper floor there are compact rooms for medical, dental, legal and psychological care plus a waiting room. Facing the street there is also an experimental kitchen that is used for training classes open to the community. There is also a small shop open to the street that offers the possibility of selling the goods that are produced there. An upper garden terrace floor houses a small kids' playroom; it is open for supervised activities such as gardening, yoga, etc. Anyway, the arrangement of the spaces is highly flexible, so nowadays, after some years of being in use, some of these activities have given room to others, according to the NGO and the community changing needs - as it was expected.

The two pavilions are connected by elevated passageways, and the rooms on the upper floors, especially those meant to be used by the community, are very open and transparent, as to signalize that everyone is welcome, and to dispel the initial mistrust that could have arisen within the community, when the Projeto Viver Association initially moved to its headquarters.

The building structure is quite simple, using reinforced concrete and concrete block walls. The metal window frames are sometimes protected by galvanized steel flap louvers, or else, with tilting metallic elements that may be closed to darken the room's interiors during day classes. The staircases and the metal walkways were made within an expanded metallic mesh system, as well as the overhead door of the workshop and the sealing perforated plates from the caretaker's house. The building construction intentionally employs simple and regular materials that are very commonly used in the poorer neighborhoods of the city, such as the concrete blocks and ceramic shards. But here these materials are disposed in a somewhat different and innovative way: the building integrates to the place, without imitating it, but seeking to propose its aesthetic requalification.

Finally, some words about the authors. This building was designed when the architects counted only 5 years of practice. In 2000 as young architects Fernando Forte, Lourenço Gimenes and Rodrigo Marcondes Ferraz founded FGMF Architects, and since then their firm has been steadily growing up. Today, fifteen years later, they have already developed 350 projects. Despite the relatively large volume of projects, theirs is still a relatively small office, looking for opportunities to experiment as much as possible with different scales, materials and demands. Such diversity helps them keep their highly investigative performance and a healthy business management focused on design excellence and high productivity. They say they believe that the critical acceptance of the error is the key to a good design process: such anti-formalistic or "erratic" behavior is perhaps what stimulates them to achieve a smart adaptability, which makes them able to cope with each different situation.

Latin America is almost a continent. It would be preposterous of me to even think about aptly representing its contemporary architecture with just one building. Anyway, I hope this quite small and low-key piece of architecture is also apt to show you something about our urban and architectural realities – as such, in the plural.

(All pictures are provided by the author)

主编简介

李翔宁

同济大学建筑与城市规划学院副院长、教授、博士生导师；哈佛大学设计研究院客座教授；2016年入选教育部"长江学者奖励计划"青年学者；国际建筑评论家委员会委员。

作为研究中国当代建筑理论、批评的知名学者，作为知名策展人，在国际和国内学术刊物发表了大量关于中国当代建筑与城市的论文，并担任当代建筑欧洲联盟奖——密斯·凡·德·罗奖、西班牙国际建筑奖等多个国际重要建筑奖项的评委。曾受邀在哈佛大学、麻省理工学院、普林斯顿大学、加拿大建筑中心等多所国际著名大学和学术机构演讲。

曼努埃尔·夸德拉

1952年生于秘鲁，利马。德国卡塞尔大学建筑、城市规划与景观学院建筑历史/建成环境历史系教授。曾在利马学习建筑学专业，并于达姆施塔特工业大学获得博士学位。此后，曾在达姆施塔特工业大学、海德堡大学、贝尔格拉诺大学（阿根廷，布宜诺斯艾利斯）、史德尔学院（德国，法兰克福）与塔尔卡大学（智利）从事教学与研究。以作家与策展人身份广泛参与活动，现为德国建筑师协会（BDA）杰出会员，国际建筑评论家委员会（CICA）正式会员。自2005年起，担任CICA秘书长，并于2008年起担任CICA指导委员会成员。2011年，获得秘鲁利马工业大学荣誉博士学位。

Brief Introduction of the Editors

Li Xiangning

As deputy dean and full professor in history, theory and criticism at Tongji University College of Architecture and Urban Planning, Li Xiangning is a member of CICA (Comité International des Critiques d'Architecture), and has published widely on contemporary Chinese architecture and urbanism in international architectural magazines including *Architectural Review, A+U, Architectural Record, Arquitectura Viva, Space, Domus,* and *Volume*.

He has been a jury member to many international awards and competitions including Mies van der Rohe Award — the European Union Prize for Contemporary Architecture and the Spanish International Architectural Prize. In 2016, he was a Visiting Professor in Architecture, teaching at Harvard GSD. He was elected Young Yangtze River Scholar in 2016. He lectured in universities and institutes including Princeton University, University of Southern California, Chalmers University, UIAV, IAAC, University of Florence, Politecnico de Milano, Canadian Center for Architecture.

Manuel Cuadra

Born in 1952 in Lima, Peru, Manuel Cuadra is a professor of architectural history / history of the built environment at the Department of Architecture, Urban Planning, and Landscape Planning at the University of Kassel (Germany). After studying architecture in Lima and becoming a doctor at the TU Darmstadt he concentrated on teaching and research at the Technical University of Darmstadt, University of Heidelberg, the University of Belgrano in Buenos Aires (Argentina), the Städel School in Frankfurt and the University of Talca in Talca (Chile).

Because of his extensive activity as a writer and a curator, Manuel Cuadra was appointed Extraordinary Member of the Association of German Architects BDA, and a full member of the International Committee of Architectural Critics (CICA) – since 2008 he serves CICA as a member of the Board of Directors and since 2005 as Secretary General. In 2011 he received an honorary doctorate from the Universidad Nacional de Ingeniería, Lima (Peru).

图书在版编目（CIP）数据

全球化进程中的地方性建筑策略：国际建筑评论家委员会研讨会论文集：汉英对照 / 李翔宁,（秘）曼努埃尔·夸德拉（Manuel Cuadra）主编. -- 上海：同济大学出版社，2019.10
ISBN 978-7-5608-8770-8

Ⅰ.①全… Ⅱ.①李… ②曼… Ⅲ.①建筑设计－世界－文集－汉、英 Ⅳ.①TU2-53

中国版本图书馆CIP数据核字(2019)第224061号

全球化进程中的地方性建筑策略
国际建筑评论家委员会研讨会论文集

李翔宁 〔秘〕曼努埃尔·夸德拉——主编

出 版 人……华春荣	
策　　划……秦蕾 / 群岛工作室	
责任编辑……李争	
责任校对……徐春莲	
平面设计……付超	
版　　次……2019年10月第1版	
印　　次……2019年10月第1次印刷	
印　　刷……上海安枫印务有限公司	
开　　本……787mm×1092mm 1/16	
印　　张……9.5	
字　　数……237 000	
书　　号……ISBN 978-7-5608-8770-8	
定　　价……78.00元	
出版发行……同济大学出版社	
地　　址……上海市杨浦区四平路1239号	
邮政编码……200092	
网　　址……http://www.tongjipress.com.cn	
经　　销……全国各地新华书店	

本书若有印装质量问题，请向本社发行部调换。
版权所有 侵权必究
光明城联系方式……info@luminocity.cn

Local Architecture Strategies in the Context of Globalization
Conference Proceedings of the International Committee of Architectural Critics (CICA)

Edited by : Li Xiangning　　[Peru] Manuel Cuadra

ISBN 978-7-5608-8770-8

Initiated by : Qin Lei / Studio Archipelago
Produced by : Hua Chunrong (publisher), Li Zheng (editing), Xu Chunlian (proofreading), Fu Chao (graphic design)
Published in October 2019, by Tongji University Press,
1239, Siping Road, Shanghai, China, 200092.
www.tongjipress.com.cn
All rights reserved
No part of this book may be reproduced in any manner whatsoever without written permission from the publisher, except in the context of reviews.
Contact us: info@luminocity.cn

本书受国家自然科学基金面上项目"基于关联域批评话语分析的当代中国建筑国际评价认知模式与传播机制研究"（51878451）资助

luminocity.cn

光 明 城

LUMINOCITY

"光明城"是同济大学出版社城市、建筑、设计专业出版品牌,由群岛工作室负责策划及出版,致力以更新的出版理念、更敏锐的视角、更积极的态度,回应今天中国城市、建筑与设计领域的问题。